Optimization of Biodiesel, Methanol and Methane Production and Air Quality Improvement

Optimization of Biodiesel, Methanol and Methane Production and Air Quality Improvement

Special Issue Editor

João Fernando Pereira Gomes

MDPI • Basel • Beijing • Wuhan • Barcelona • Belgrade

Special Issue Editor
João Fernando Pereira Gomes
Department of Chemical
Engineering (ADEQ),
Instituto Superior de
Engenharia de Lisboa (ISEL),
R. Conselheiro Emídio Navarro
Portugal

Editorial Office
MDPI
St. Alban-Anlage 66
4052 Basel, Switzerland

This is a reprint of articles from the Special Issue published online in the open access journal *Energies* (ISSN 1996-1073) in 2019 (available at: https://www.mdpi.com/journal/energies/special_issues/ Biodiesel_Methanol_Methane).

For citation purposes, cite each article independently as indicated on the article page online and as indicated below:

LastName, A.A.; LastName, B.B.; LastName, C.C. Article Title. *Journal Name* **Year**, *Article Number*, Page Range.

ISBN 978-3-03928-100-8 (Pbk)
ISBN 978-3-03928-101-5 (PDF)

© 2019 by the authors. Articles in this book are Open Access and distributed under the Creative Commons Attribution (CC BY) license, which allows users to download, copy and build upon published articles, as long as the author and publisher are properly credited, which ensures maximum dissemination and a wider impact of our publications.
The book as a whole is distributed by MDPI under the terms and conditions of the Creative Commons license CC BY-NC-ND.

Contents

About the Special Issue Editor vii

Preface to "Optimization of Biodiesel, Methanol and Methane Production and Air Quality Improvement" ix

Sanghyun Park, Yongtae Ahn, Young-Tae Park, Min-Kyu Ji and Jaeyoung Choi
The Effect of Mixed Wastewaters on the Biomass Production and Biochemical Content of Microalgae
Reprinted from: *Energies* **2019**, *12*, 3431, doi:10.3390/en12183431 1

Sanghyun Park, Yongtae Ahn, Kalimuthu Pandi, Min-Kyu Ji, Hyun-Shik Yun and Jae-Young Choi
Microalgae Cultivation in Pilot Scale for Biomass Production Using Exhaust Gas from Thermal Power Plants
Reprinted from: *Energies* **2019**, *12*, 3497, doi:10.3390/en12183497 14

Ana Gonçalves, Jaime Filipe Puna, Luís Guerra, José Campos Rodrigues, João Fernando Gomes, Maria Teresa Santos and Diogo Alves
Towards the Development of Syngas/Biomethane Electrolytic Production, Using Liquefied Biomass and Heterogeneous Catalyst
Reprinted from: *Energies* **2019**, *12*, 3787, doi:10.3390/en12193787 24

Shemelis N. Gebremariam, Trine Hvoslef-Eide, Meseret T. Terfa and Jorge M. Marchetti
Techno-Economic Performance of Different Technological Based Bio-Refineries for Biofuel Production
Reprinted from: *Energies* **2019**, *12*, 3916, doi:10.3390/en12203916 45

S. Ozkan, J. F. Puna, J. F. Gomes, T. Cabrita, J. V. Palmeira and M. T. Santos
Preliminary Study on the Use of Biodiesel Obtained from Waste Vegetable Oils for Blending with Hydrotreated Kerosene Fossil Fuel Using Calcium Oxide (CaO) from Natural Waste Materials as Heterogeneous Catalyst
Reprinted from: *Energies* **2019**, *12*, 4306, doi:10.3390/en12224306 66

Marta Ramos, Ana Paula Soares Dias, Jaime Filipe Puna, João Gomes and João Carlos Bordado
Biodiesel Production Processes and Sustainable Raw Materials
Reprinted from: *Energies* **2019**, *12*, 4408, doi:10.3390/en12234408 85

Samuel Santos, Luís Nobre, João Gomes, Jaime Puna, Rosa Quinta-Ferreira and João Bordado
Soybean Oil Transesterification for Biodiesel Production with Micro-Structured Calcium Oxide (CaO) from Natural Waste Materials as a Heterogeneous Catalyst
Reprinted from: *Energies* **2019**, *12*, 4670, doi:10.3390/en12244670 115

About the Special Issue Editor

João Gomes is a habilitated Professor of Chemical Technology at ISEL, Lisbon Polytechnic, Portugal, where he is the responsible for the MSc Course on Chemical and Biological Engineering, and a researcher of CERENA—Center for Natural Resources and Environment from Lisbon University. He holds a BSc and a PhD in Chemical Engineering both from the Technical University of Lisbon. He has published more than 100 papers in scientific referred journals and 22 books, and has managed several international R&D projects. His current research interests include clean chemical processes; clean energy and fuels; heterogeneous catalysis; nanotoxicology and particulate toxicology; monitoring and control of air pollution; CO2 capture.

Preface to "Optimization of Biodiesel, Methanol and Methane Production and Air Quality Improvement"

Alternative and renewable energy sources already play a very decisive role in the development of human society, helping to fulfill increasing energy demands from both industrialized and underdeveloped countries, as well as economic needs, which must comply with a decarbonized economy, decreasing the energy impact on the global environment. Among these alternative energy sources, fuels such as biodiesel, methanol, and methane are good examples of how the previous design can be achieved, as these fuels can be obtained from renewable sources, used in applications such as transportation systems, electricity generation, fuel conversion, and even for electricity storage, with reduced impact on air emissions. Although a great deal of research has been done, and there have been important advances, there is still a need to increase research, particularly in further development of new approaches, their optimization and practical compatibilization with existing systems, so that these new technologies can be efficiently adopted by energy stakeholders, while also being cost-competitive and truly effective.

João Fernando Pereira Gomes
Special Issue Editor

Article

The Effect of Mixed Wastewaters on the Biomass Production and Biochemical Content of Microalgae

Sanghyun Park [1,2,†], Yongtae Ahn [1,†], Young-Tae Park [3] , Min-Kyu Ji [4] and Jaeyoung Choi [1,*]

1. Center for Environment, Health and Welfare Research, Korea Institute of Science and Technology, Seoul 02792, Korea
2. Graduate School of Energy and Environment (KU-KIST GREEN SCHOOL), Korea University, Seoul 02841, Korea
3. Natural Products Research Institute, Korea Institute of Science and Technology, Seoul 02792, Korea
4. Environmental Assessment Group, Korea Environment Institute, Sejong 30147, Korea
* Correspondence: jchoi@kist.re.kr; Tel.: +82-2-958-5846
† These authors contributed equally to the paper.

Received: 31 July 2019; Accepted: 3 September 2019; Published: 5 September 2019

Abstract: The effect of ammonia and iron concentration in Bold Basal Medium and mixed wastewater (including pretreated piggery wastewater and acid mine drainage) on biomass production and biochemical content (lipid and ß-carotene) of microalgae (*Uronema* sp. KGE 3) was investigated. Addition of iron to the Bold Basal Medium enhanced the growth, lipid, and ß-carotene of *Uronema* sp. KGE 3. The highest dry cell weight, lipid content, and lipid productivity of KGE 3 were 0.551 g L^{-1}, 46% and 0.249 g L^{-1} d^{-1}, respectively, at 15 mg L^{-1} of Fe. The highest ß-carotene was obtained at 30 mg L^{-1} of Fe. The biomass production of KGE 3 was ranged between 0.18 to 0.37 g L^{-1}. The microalgal growth was significantly improved by addition of acid mine drainage to pretreated piggery wastewater by membrane. The highest dry cell weight of 0.51 g L^{-1} was obtained at 1:9 of pretreated piggery wastewater by membrane and acid mine drainage for KGE 3. The removal efficiencies of total nitrogen and total phosphate was ranged from 20 to 100%. The highest lipid and ß-carotene content was found to be 1:9. Application of this system to wastewater treatment plant could provide cost effective technology for the microalgae-based industries and biofuel production field, and also provide the recycling way for pretreated piggery wastewater and acid mine drainage.

Keywords: microalgae; acid mine drainage; microalgae culture; lipid; ß-carotene

---

1. Introduction

Microalgae synthesize contain valuable biomass compounds, such as lipid, ß-carotene, astaxanthin, and lutein [1,2]. Biofuel has been obviously reported as a source of renewable energy for superseding the fossil fuel [3]. The biodiesel can be derived from lipids of microalgae, which is achievable since microalgae contains up to 50% of lipid with respect to dry cell weight [4]. Also, microalgae have advantages like high photosynthetic efficiency, high productivity yield, non-arable land use, and ability to capture and utilize CO_2 as a nutrient [5,6]. Carotenoids are commercially using as food coloring agent and feed cosmetic products, particularly ß-carotene acted as an antioxidant, anticancer and immune functions. The nutrient concentration is one of the vital sources for successful of cultivation microalgal, and efficient and improved synthesis of microalgae biomass for subsequent production of biodiesel and ß -carotene has also been reported [7]. Also, the concern has been increasing towards the biomass production from microbial to absorb heavy metal ions and bioaccumulation [8,9].

Acid mine drainage (AMD) is often net acidic and contains a high concentration of essential metals including iron, which can be used as micronutrients to improve the growth of microalgal with respect to enhancing the lipid production [10,11]. Organic sources for microalgae growth was accounted as 80% of the total cost of the medium [12], hence it is necessary to use low-cost organic moieties or wastewaters to achieve maximum microalgal yield at commercial scale [13]. The development of livestock production has environmental problems. Manure characterization results show that it contains a high concentration of organic material like nitrogen and phosphorus. The manure treatment process tends to couple with anaerobic digestion, which reduces the treatment cost along with nutrient recovery for microalgae, and manure then becomes a resource [14]. The different type of wastewaters from municipal, piggery and food industries was used as resource for microalgae cultivation and its growth has been investigated [15,16].

The mixing of different wastewaters to acquire the ideal nutrients ratio for improved biomass production has not been well investigated. In addition to that, supplementation of manure wastewater (as source of TN and TP) with AMD (as source of Fe) has not yet been studied. Hence, the present investigation to explore the effect of mixed wastewater (including manure and AMD) on biomass production, biochemical content (lipid and ß-carotene) of microalgae (Uronema sp. KGE 3) was examined. The removal of total nitrogen (TN) and total phosphorus (TP) was monitored. The kinetic assessment of specific growth rate of microalgae was also evaluated. These results provide fundamental information to establish and cultivate the microalgae in Pipes inserted microalgae reactor (PIMR) system, which could be one of the alternate choices for a microalga-based biodiesel production strategy.

2. Materials and Methods

2.1. Preparation of Microalgal Suspension, Strain, and Culture Medium

Uronema sp. KGE 3 microalgae was derived from AMD at YD abandoned mine (Gangwon-do, South Korea). The resultant microalga was cultivated in 2.2 L column using 5% ($V_{inoculum}/V_{media}$) of 2.0 L of Bold's Basal Medium (BBM) [17]. The cultivated Uronema sp. KGE 3 microalgae was identified by rDNA D1-D2 sequence-based phylogenetic extraction analysis (Figure 1). The cultures were kept under a light irradiation of 130 µmol photon m^{-2} s^{-1} at 25 ± 1 °C for ten days with white fluorescent light illumination. Throughout the incubation, each column was constantly sparged with air through 0.2 µm sterilized filter at the flow rate value of 0.4 L min^{-1}, the suspension stirred, and concurrently supplied with atmospheric CO_2. The microalga biomass suspension in the medium, which adjusted to an absorbance of 0.01 at an optical density of 680 nm as measured by a spectrophotometer (Hach DR/2800, Loveland, Colorado, USA). Experiments were used stock microalgae to inoculum amount of three milliliters.

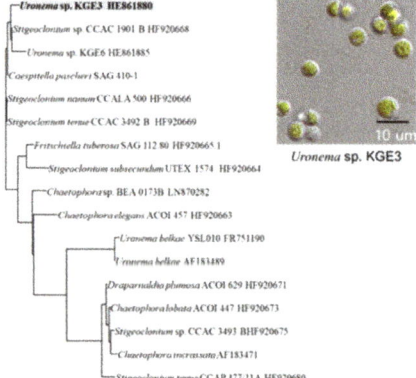

Figure 1. Uronema sp. KGE3 cell morphology and phylogenetic tree displays the connection between the NSU(Nuclear large subunit) rDNA D1-D2 sequences of the isolates Uronema sp. KGE3 and the maximum parallel sequences rescued from the NCBI nucleotide database.

2.2. Wastewater Sampling and Analysis

Acid mine drainage (AMD) was collected from Yeong-dong mine at Gangneung, South Korea. The manure was collected from cattle farms at Jeollanam-do Province, South Korea. The manure was pretreated by anaerobic digestion and membrane bioreactor (MBR). Wastewater was immediately filtered using syringe filter (Polyvinylidene fluoride or polyvinylidene difluoride (PVDF), 0.2 μm) to remove the suspended solid particles [18]. Total nitrogen (TN), total phosphorus (TP), and ammonium was measured using Hach Kit (HACH, CO, USA), which are equivalent methods: Standard Methods 4500-N C and 4500-P B (5). Metal ions in AMD were analyzed by inductively coupled plasma optical emission spectroscopy (ICP-OES, Varian 730-ES, USA). The Orion 5-Star pH/DO/ORP/Cond. Meter (Thermo Scientific, USA) was used to measure the solution pH. The major components of AMD were 237.8 mg L^{-1} of total iron, 187.3 mg L^{-1} of ferrous, 5.7 mg L^{-1} of manganese and 320.4 mg L^{-1} of sulfate, and solution pH was 3.2. The major components present in the pretreated piggery wastewater by membrane (PPWM) were 4645 mg L^{-1} of ammonium, 6230 mg L^{-1} of total nitrogen, and 365 mg L^{-1} of total phosphate, and solution pH was found to be 8.53.

2.3. Experimental Method

The experiments were conducted in two stages. In the first stage, BBM was supplemented with different concentration of iron and ammonium to optimize the concentration for further experimental. The second phase was the replacement of BBM with PM amended with AMD as a cheap source of iron. The mixed ratio of PPWM with AMD were 1:0, 1:1, 1:9, and 1:19. The biomass production, nutrients removal (including TN /TP), and biochemical composition (including lipids and ß-carotene) of the microalga in all tested condition were evaluated.

Growth rate was used to monitor the microalgae by absorbance (optical density (OD)) and gravity (dry cell weight (DCW)) method. OD was measured at 680 nm using a spectrophotometer (HS-3300; Humas, Daejeon, Korea). To measure the DCW, microalgae was harvested by centrifugation at 3000 rpm for 20 mins (1580R, Labogene, Korea). The harvested microalgal biomass was frozen at −70 °C by deep freezer (DFU-256CE, Operon, Korea). After freeze-dried the biomass, experiments sample were lyophilized by freezing dryer (FDB-550, Operon, Korea).

The DCW was used to calculate the microalgae growth by: biomass productivity (P), as demonstrated in Equation (1).

$$P = M_b - M_{b0}/T - T_0 \tag{1}$$

where M_b and M_{b0} are microalgae dry biomass at time T and initial time T_0. The specific growth rate (μ, day $^{-1}$) of the microalgae can be calculated using Equation (2) as follows [19].

$$\mu = \frac{\ln X_i - \ln X_0}{t_i - t_0} \tag{2}$$

2.4. Analysis of Lipid and ß-carotene

The dried biomass of freeze-dried microalgae was used to extract the total lipids contents by modified method reported by Bligh and Dyer [20]. After the extraction, chloroform-methanol layer contains the microalgae extracted lipids, which can be collected using a separate-funnel and followed by the rotary evaporator. The lipid content of microalgae was calculated by the following Equation (3):

$$\text{Lipid content (\%)} = (W_2 - W_3)/W_1 \times 100 \tag{3}$$

where W_1 is weight of dried cell biomass, W_2 is glass tube with extracted lipids W_3 is empty glass tube.

Various organics solvents including methanol, acetone, and chloroform were used to extract the ß-carotene from freeze-dried algal biomass. In this study, the extraction of ß-carotene from freeze-dried algal biomass using chloroform provides a quietly high amount of yield. Hence, 0.2 g of freeze-dried algal biomass was suspended in 5.0 mL of chloroform, then sonicated for 10 min under maximum power in an ultrasound condition, and finally the suspension was kept at 4 °C for 24 h. The supernatant liquid in the suspension was harvested by centrifugation and stored until analysis was carried out. After 14 extractions, chloroform did not display any coloration while the pellet remained greenish in color. The ß-carotene was analyzed by LC-MS (e2695, Waters, USA) and analytical column with 5 mm C30-reversed phase material (250 mm L 4.6 mm ID) at 30 °C. Detail of pretreatment and analyze procedure followed the Gupta method [21].

3. Results and Discussion

3.1. The Influence of Iron and Ammonium on the Microalgal Growth

The impact of iron on the growth of lipid, ß-carotene, and *Uronema* sp. KGE 3 cultivated in BBM are appeared in Figure 2A–C, respectively. The Fe concentration used in this study was 5 mg L^{-1} (from BBM only), 10, 15, 25, and 35 mg L^{-1}. Fe acted as a vital element for the growth of higher plants and microorganisms including microalgae [22]. The growth and reproduction of microalgae also requires addition of Fe. In microalgae-based reactors, optimal iron dosage needs to be maintained for stable growth of microalgae biomass [23]. The impact of various concentration of Fe on the growth and lipid production of *Uronema* sp. KGE 3 was investigated to identify the required amount of Fe from AMD. Addition of iron (15-35 mg L^{-1}) to BBM enhanced the growth and biochemical content (including lipid and ß-carotene) of *Uronema* sp. KGE 3. The highest dry cell weight (Figure 2A), lipid content, and lipid productivity (Figure 2B) of KGE 3 were 0.551 g L^{-1}, 46%, and 0.249 g L^{-1} d^{-1}, respectively after the addition of 15 mg L^{-1} of Fe, and it was higher than control by 24% (Figure 2A). The highest ß-carotene was found at 35 mg L^{-1} of Fe (Figure 2C). The metals including Fe, Mn, Cu, Zn, and Co acted as micronutrients and possessed the ability to increase the growth of microalgae by certain concentration, while on the other hand, the higher concentrations of these metals can hinder the growth of microalgae [24,25]. The growth pattern of *Uronema* sp. KGE 3 cultivated in BBM under various concentration of ammonium is shown in Figure 3A–C. Ammonium concentration used in this study was 0 (BBM only) to 1,000 mg L^{-1}. The growth of KGE 3 was ranged between 0.18 to 0.37 g L^{-1} (Figure 3A). The growth pattern of KGE 3 under ammonium was below the control (Figure 3B), which revealed the toxicity of ammonium. The present result was reliable with previous report displaying that ammonium had strong effect on the microalgal growth [26]. The highest lipid and ß-carotene content was achieved at 1000 mg L^{-1} of ammonium (Figure 3C).

The microalgal with high specific growth rate (μ) is important criteria to select the best microalgal species as it signifies a shorter doubling time [27]. The changes on the specific growth rate of *Uronema* sp. KGE 3 cultivated under the effect of iron and ammonium for 9 days in BBM is shown in Figure 4A,B, respectively. The highest specific growth rate of *Uronema* sp. KGE 3 was observed at 15 and 25 mg L^{-1} of Fe during the cultivation time (Figure 4A), while the trended specific growth rate under ammonium was not fixed, which might be due to the pernicious effect of ammonium (Figure 4B). The changes on specific growth rate was observed in this research because of the difference in specific cell biomass yield [28]. Yoshimura et al. reported that *Botryococcus braunii* possesses the specific growth rate and doubling time of 0.19–0.50 day^{-1} and 1.4–3.6 days, respectively [29]. Zhu et al. utilized artificial wastewater in photobioreactor for both cultivation and treatment of wastewater using *Chlorella zofingiensis*. The result shows that wastewater treatment was ranged from 0.208 to 0.260 day^{-1} with a doubling time of 2.67 to 3.34 days within 15 days of cultivation [30]. The difference in values of specific growth rate and doubling time for *Chlorella zofingiensis* might be owing to variances in cultivation period, culture medium, and cultivation conditions [31].

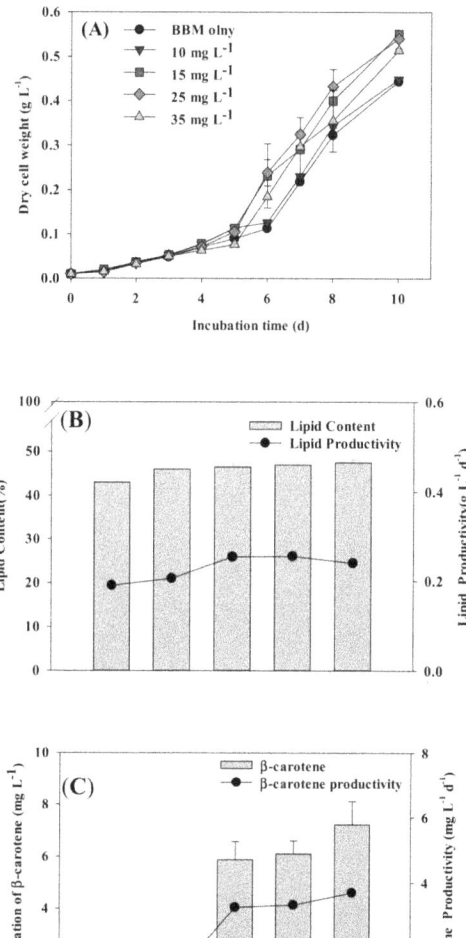

Figure 2. The effect of iron on the growth (**A**), lipid (**B**) and ß-carotene (**C**) of *Uronema* sp. KGE 3 cultivated in Bold's Basal Medium (BBM).

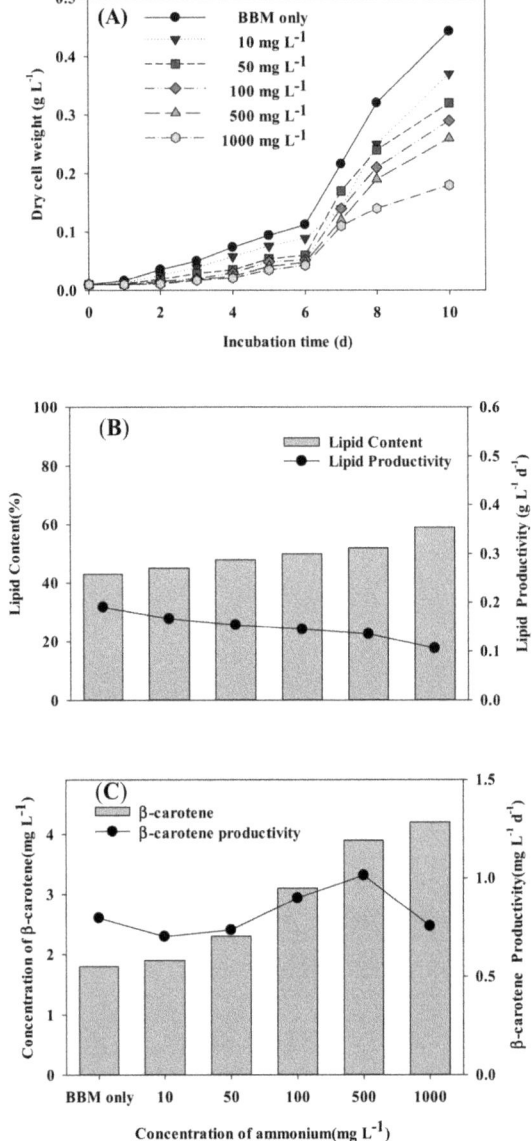

Figure 3. The effect of ammonium the growth (**A**), lipid (**B**) and ß-carotene (**C**) of *Uronema* sp. KGE 3 cultivated in Bold's Basal Medium (BBM).

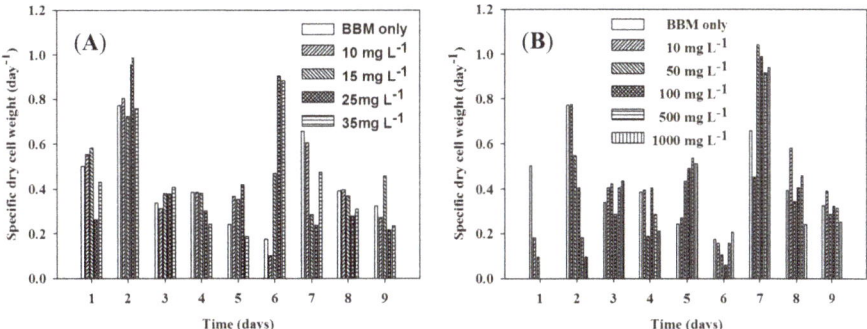

Figure 4. Changes on the specific growth rate of *Uronema* sp. KGE 3 under the effect of iron (**A**) and ammonium (**B**) cultivated for 9 days in Bold's Basal Medium (BBM).

3.2. Effect of Pretreated Piggery Wastewater by Membrane (PPWM) and Acid Mine Drainage (AMD) on the Microalgal Growth

The growth of *Uronema* sp. KGE 3 in various ratio of PPWM present of AMD is presented in Figure 5. The physico-chemical characteristics of experimental variations are presented in Table 1. The suitable selection of wastewater, robust microalgal species, and pre-treatment techniques are the primary factors for microalgae-based biofuel production, as well as the development of technology with the combination of wastewater treatment and microalgae cultivation for biomass and lipid production [32,33]. The dry cell weight of microalgae ranged between 0.19 to 0.51 g L^{-1} in all the tested conditions. The microalgal growth was significantly improved by addition of AMD to PPWM. The highest dry cell weight (0.51) was obtained at 1:9 of PPWM and AMD for *Uronema* sp. KGE 3 (Figure 5). Pervious studies have reported that the addition of AMDs to the microalgal culture enhanced the microalgal growth due to AMDs provided an optimal C:N:P ratio and suitable initial pH [18]. Microalgae can grow copiously when provide the adequate amount of nutrients and appropriate conditions. The algal growth is also directly influenced by the temperature, nutrients (TN, TP), light intensity, iron concentration, and the initial pH level [34,35]. Addition of AMD in this study provided suitable pH, which resulted in a higher growth rate than without addition of AMD (Figure 5A). The initial pH level of microalgal culturing media (including Bold's basal medium) was 6.8 [36].

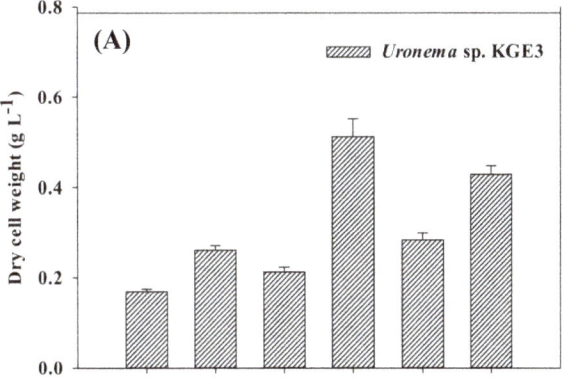

Figure 5. *Cont.*

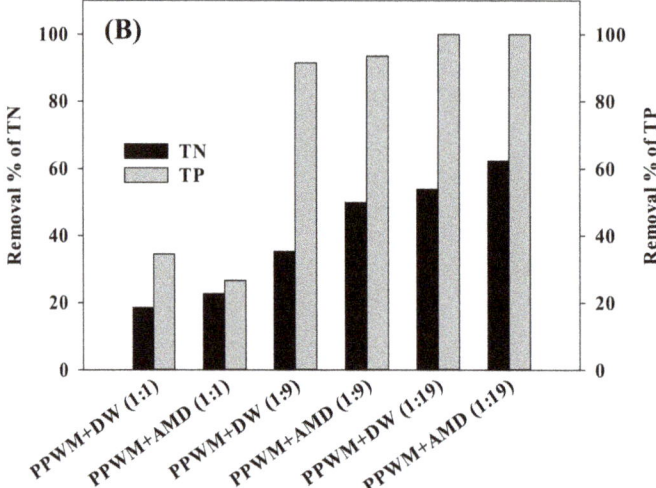

Figure 5. The effect of various mixing ratios between pretreated piggery wastewater by membrane (PPWM) and acid mine drainage (AMD) on the growth (**A**) and nutrients removal (**B**) of *Uronema* sp. KGE 3.

Table 1. Experimental condition for studying the effect of mixture from Pretreated manure and acid mine drainage.

Parameters	Experimental Condition	Pretreated Manure + Distilled Water (v/v)			Pretreated Manure + Acid Mine Drainage (v/v)		
		1/1	1/9	1/19	1/1	1/9	1/19
Total nitrogen (mg L^{-1})		743	173	76	762	168	69
Ammonium (mg L^{-1})		726	162	79	732	148	69
Nitrate (mg L^{-1})		13	3	1	15	2	1
Total phosphorous (mg L^{-1})		29	5.9	2.9	30	6.2	3.1
Manganese (mg L^{-1})		-	-	-	2.9	0.6	0.3
Sulfate (mg L^{-1})		-	-	-	160.2	32	16
Iron (mg L^{-1})		4.6	0.9	0.5	39.6	69.8	73.7

3.3. Effect of Microalgal Growth on Nutrient Removal

The microalgae growth can be enhanced by utilization of trace elements and nutrients present in the wastewater and subsequently the nutrient concentration in the wastewater reduced, which supports the advanced of wastewater recycling process and biomass production. Aliquots of wastewater samples were collected after 10 days of cultivation period to calculate the TN and TP removal from various dilutions of PPWM with AMD by *Uronema* sp. KGE 3, in order to assess its efficiency of nutrient removal. The removal efficiency was ranged from 18.6 to 62.3% for TN, while TP was ranged from 26.7 to 100% (Figure 5B). Nitrogen is one of the main macro elements for microalgae growth, which ranges from 1.0 to 10.0% of total dry biomass and is also an indicator for determining lipid content in microalgae biomass [37]. The TP was not fully used as nutrient by microalgae. Only part of the TP was utilized as a nutrient for growth and metabolism, including energy transfer and the biosynthesis of DNA, while the remaining TP was precipitated and assimilated into biomass through intracellular polyphosphate. Schreiber et al., reported that most of the consumed P accumulated in their body [38].

3.4. The Production of Lipid and ß-carotene after Cultivation in Pretreated Piggery Wastewater by Membrane (PPWM) Supplemented with Acid Mine Drainage (AMD)

After cultivation of *Uronema* sp. KGE 3 in PPWM supplemented with AMD, lipid and ß-carotene content was evaluated (Figure 6). The lipid content was ranged between 36 to 52% (Figure 6A), while ß-carotene content was ranged between 0.5 to 5.9% (Figure 6B). The highest lipid and ß-carotene content was found at 1:9 (PPWM: AMD ratio) medium, when *Uronema* sp. KGE 3 cultivated in BBM showed an optimal growth and obtained a higher cell growth than 1:9 (PPWM: AMD ratio) medium. Even though, the value of DCW in BBM (1.04 g L^{-1}) was higher than PPWM: AMD medium (0.91 g L^{-1}), 1:9 (PPWM: AMD ratio) medium was more suitable media for productivity of lipid and ß-carotene. The result indicated that iron in AMD could promote the biomass productivity (Figure 7). The addition of PPWM (as a source of nutrients) and AMD to the system increases the microalgal biomass from 0.19 to 0.51 g L^{-1} and also increases the lipid/carotenoids productivity from 0.166 g L^{-1} d^{-1} and 1.01 mg g^{-1} d^{-1} to 0.251 g L^{-1} d^{-1} and 3.01 mg g^{-1} d^{-1} respectively, along with increasing nutrient removal efficiencies from 18.6 to 62.3%.

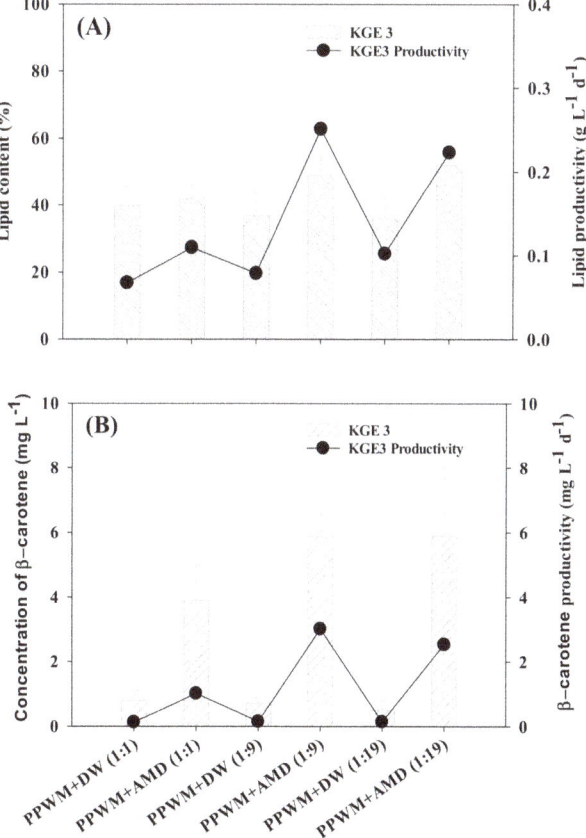

Figure 6. Effect of various mixing ratios between pretreated piggery wastewater by membrane (PPWM) and acid mine drainage (AMD) on the lipid production (**A**) and ß-carotene (**B**) of *Uronema* sp. KGE 3.

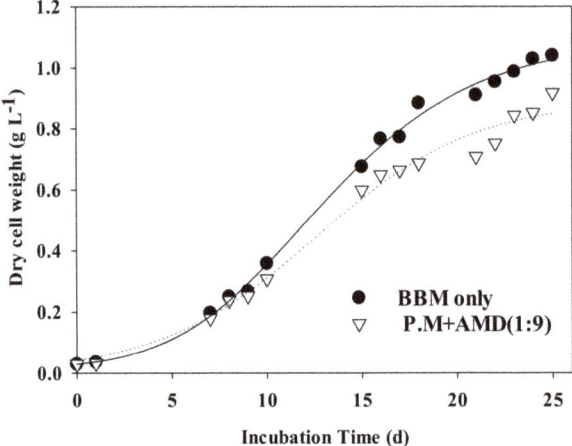

Figure 7. Time courses of biomass concentration for *Uronema* sp. KGE 3 cultivated in Bold's Basal Medium (BBM). with pretreated piggery wastewater by membrane (PPWM) + acid mine drainage (AMD) (1:9) medium during Pipes inserted microalgae reactor (PIMR) operation for 25 days.

3.5. Comparison of Other Researches for Lipid and ß-carotene Production

Iron may affect the productivity of lipid content, while ammonia affects both lipid and β-carotene contents. Table 2 lists the microalgae production capacities of lipid and β-carotene. Kim et al. (2007) [39] reported that fermented swine urine was used as source of ammonia. It is worth noticing that while increasing the amount of ammonia, the ß-carotene production was also increased. In this study, ß-carotene production is well related to the concentration of PPWM, because PPWM acts as an ammonia source which correspondingly increases the production of ß-carotene. Feng et al. (2011) [40] also added ammonium in artificial wastewater to enhance the yield of ß-carotene. Singh et al. (2015) [41] reported that lipid productivity was 3–5 times higher than the conventional media.

Table 2. Comparison of other researches for lipid and ß-carotene production.

Strain	Cultivation Condition	Lipid (g/L d)	β-carotene (mg/g d)	Reference
Uronema sp. KGE 3[a]	Iron	0.249	3.69	This study
Uronema sp. KGE 3[a]	Ammonia	0.166	1.01	This study
Scenedesmus spp. -Mixed culture[b]	Fermented swine urine	-	0.05	Kim et al. (2007) [39]
C. vulgaris FACHB1068[c]	Artificial wastewater	0.147	-	Feng et al. (2011) [40]
Ankistrodesmus falcatus KJ671624	Iron	0.074	-	Singh et al. (2015) [41]
Uronema sp. KGE 3[a]	AMDS+PPWM	0.251	3.01	This study

[a] This study; [b] Kim et al. (2007); [c] Feng et al. (2011); [d] Singh et al. (2015).

4. Conclusions

The condition of PPWM: AMD = 1:9 and PPWM: AMD = 1:19 leads to the production of microalga feedstock and is suitable for obtaining highly efficient lipid and ß-carotene. Utilization of *Uronema* sp. KGE 3 as microalgae for the generation of biofuel material and recycling of wastewater could be a cost-effective and environmentally sustainable strategy. The optimal iron concentration of 15 mg L^{-1}

has potential application for microalgae cultivation because of distribution of biomass productivity increase. The change in iron and ammonium concentration can induce the lipid and ß-carotene accumulation. A pilot-scale study using PIMR proved that PPWM 1: AMD 9 can be used as effective cultivation media instead of synthetic BBM. This culture approach and appropriate microalgae strains will provide the cost-effective technology to microalgae-based industries and biofuel production fields and a recyclable way for PPWM and AMD. Application of this approach to pig wastewater treatment system could be an alternative strategy to reduce contaminant concentration such as TN and TP, and to increase facility for bioenergy production.

Author Contributions: Conceptualization, S.P. and Y.-T.P.; Methodology, S.P. and Y.-T.P.; Software, S.P.; Validation, Y.-T.P.; Investigation, S.P. and Y.-T.P.; Data Curation, S.P. and Y.-T.P.; Writing-Original Draft Preparation, S.P.; Writing-Review & Editing, Y.A. and M.-K.J.; Visualization, Y.-T.P.; Supervision, J.C.; Project Administration, J.C.; Funding Acquisition, J.C.

Funding: This work was supported by the Korea CCS R&D Center (Korea CCS 2020 Project) grant funded by the Korea government (Ministry of Science, ICT & Future Planning) in 2019 (KCRC-(2014M1A8A1049293) and the KIST Institutional Program (Project No. 2E29670).

Conflicts of Interest: The authors declare no conflict of interest.

References

1. Sigamani, S.; Ramamurthy, D.; Natarajan, H. A review on potential biotechnological applications of microalgae. *J. Appl. Pharm. Sci.* **2016**, *6*, 179–184. [CrossRef]
2. Del Campo, J.A.; Garcia-Gonzalez, M.; Guerrero, M.G. Outdoor cultivation of microalgae for carotenoid production: Current state and perspectives. *Appl. Microbiol. Biotechnol.* **2007**, *74*, 1163–1174. [CrossRef] [PubMed]
3. Crutzen, P.J.; Mosier, A.R.; Smith, K.A.; Winiwarter, W. N2O release from agro-biofuel production negates global warming reduction by replacing fossil fuels. *Atmos. Chem. Phys.* **2008**, *8*, 389–395. [CrossRef]
4. Xiong, W.; Li, X.; Xiang, J.; Wu, Q. High-density fermentation of microalga Chlorella protothecoides in bioreactor for micobio-diesel production. *Appl. Microbiol. Biotechnol.* **2008**, *78*, 29–36. [CrossRef]
5. Wang, B.; Li, Y.Q.; Wu, N.; Lan, C.Q. CO2 bio-mitigation using microalgae. *Appl. Microbiol. Biotechnol.* **2008**, *79*, 707–718. [CrossRef]
6. Biller, P.; Ross, A.B. Potential yields and properties of oil from the hydrothermal liquefaction of microalgae with different biochemical content. *Bioresour. Technol.* **2011**, *102*, 215–225. [CrossRef] [PubMed]
7. Wu, Z.; Duangmanee, P.; Juntawong, N.; Ma, C. The effects of light, temperature, and nutrition on growth and pigment accumulation of three Dunalliella salina strains isolated from saline soil. *Jundishapur J. Microbiol.* **2016**, *9*, e26732. [CrossRef]
8. Benmalek, Y.; Halouane, A.; Hacene, H.; Fardeau, M.L. Resistance to heavy metals and bioaccumulation of lead and zinc by Chryseobacterium solincola strain 1YB–R12T isolated from soil. *Int. J. Environ. Eng.* **2014**, *6*, 68–77. [CrossRef]
9. Harguinteguy, C.A.; Cirelli, A.F.; Pignata, M.L. Heavy metal accumulation in leaves of aquatic plant Stuckenia filiformis and its relationship with sediment and water in the Suquia river (Argentina). *Microchem. J.* **2014**, *114*, 111–118. [CrossRef]
10. Das, B.K.; Roy, A.; Koschorreck, M.; Mandal, S.M.; Wendt-Potthoff, K.; Bhattacharya, J. Occurrence and role of algae and fungi in acid mine drainage environment with special reference to metals and sulfate immobilization. *Water Res.* **2009**, *43*, 883–894. [CrossRef]
11. Park, Y.T.; Lee, H.; Yun, H.S.; Song, K.G.; Yeom, S.H.; Choi, J. Removal of metal from acid mine drainage using a hybrid system including a pipes inserted microalgae reactor. *Bioresour. Technol.* **2013**, *150*, 242–248. [CrossRef] [PubMed]
12. Bhatnagar, A.; Chinnasamy, S.; Singh, M.; Das, K.C. Renewable biomass production by mixotrophic algae in the presence of various carbon sources and wastewaters. *Appl. Energ.* **2011**, *88*, 3425–3431. [CrossRef]
13. Liang, Y.; Sarkany, N.; Cui, Y. Biomass and lipid productivities of Chlorella vulgaris under autotrophic, heterotrophic and mixotrophic growth conditions. *Biotechnol. Lett.* **2009**, *31*, 1043–1049. [CrossRef] [PubMed]

14. Ledda, C.; Schievano, A.; Scaglia, B.; Rossoni, M.; Fernandez, F.G.A.; Adani, F. Integration of microalgae production with anaerobic digestion of dairy cattle manure: An overall mass and energy balance of the process. *J. Clean. Prod.* **2016**, *112*, 103–112. [CrossRef]
15. Su, Y.Y.; Mennerich, A.; Urban, B. The long-term effects of wall attached microalgal biofilm on algae-based wastewater treatment. *Bioresour. Technol.* **2016**, *218*, 1249–1252. [CrossRef] [PubMed]
16. Matamoros, V.; Uggetti, E.; Garcia, J.; Bayona, J.M. Assessment of the mechanisms involved in the removal of emerging contaminants by microalgae from wastewater: A laboratory scale study. *J. Hazard. Mater.* **2016**, *301*, 197–205. [CrossRef] [PubMed]
17. Bischoff, H.W.; Bold, H.C. Phycological studies IV. Some soil algae from enchanted rock and related algal species. *Univ. Tex. Publ.* **1963**, *6318*, 1–95.
18. Ji, M.K.; Kabra, A.N.; Salama, E.S.; Roh, H.S.; Kim, J.R.; Lee, D.S.; Jeon, B.H. Effect of mine wastewater on nutrient removal and lipid production by a green microalga Micratinium reisseri from concentrated municipal wastewater. *Bioresour. Technol.* **2014**, *157*, 84–90. [CrossRef]
19. Wang, M.; Kuo-Dahab, W.C.; Dolan, S.; Park, C. Kinetics of nutrient removal and expression of extracellular polymeric substances of the microalgae, *Chlorella* sp. and *Micractinium* sp., in wastewater treatment. *Bioresour. Technol.* **2014**, *154*, 131–137. [CrossRef]
20. Bligh, E.G.; Dyer, W.J. A rapid method of total lipid extraction and purification. *Can. J. Biochem. Phys.* **1959**, *37*, 911–917. [CrossRef]
21. Gupta, P.; Sreelakshmi, Y.; Sharma, R. A rapid and sensitive method for determination of carotenoids in plant tissues by high performance liquid chromatography. *Plant Methods* **2015**, *11*. [CrossRef] [PubMed]
22. Wang, C.; Wang, X.; Wang, P.F.; Chen, B.; Hou, J.; Qian, J.; Yang, Y.Y. Effects of iron on growth, antioxidant enzyme activity, bound extracellular polymeric substances and microcystin production of Microcystis aeruginosa FACHB-905. *Ecotoxicol. Environ. Saf.* **2016**, *132*, 231–239. [CrossRef] [PubMed]
23. Zhao, Z.M.; Song, X.S.; Wang, W.; Xiao, Y.P.; Gong, Z.J.; Wang, Y.H.; Zhao, Y.F.; Chen, Y.; Mei, M.Y. Influences of iron and calcium carbonate on wastewater treatment performances of algae based reactors. *Bioresour. Technol.* **2016**, *216*, 1–11. [CrossRef] [PubMed]
24. Schowanek, D.; McAvoy, D.; Versteeg, D.; Hanstveit, A. Effects of nutrient trace metal speciation on algal growth in the presence of the chelator [S, S]-EDDS. *Aquat. Toxicol.* **1996**, *36*, 253–275. [CrossRef]
25. Miazek, K.; Iwanek, W.; Remacle, C.; Richel, A.; Goffin, D. Effect of metals, metalloids and metallic nanoparticles on microalgae growth and industrial product biosynthesis: A review. *Int. J. Mol. Sci.* **2015**, *16*, 23929–23969. [CrossRef] [PubMed]
26. Tan, X.B.; Zhang, Y.L.; Yang, L.B.; Chu, H.Q.; Guo, J. Outdoor cultures of Chlorella pyrenoidosa in the effluent of anaerobically digested activated sludge: The effects of pH and free ammonia. *Bioresour. Technol.* **2016**, *200*, 606–615. [CrossRef]
27. Vidyashankar, S.; VenuGopal, K.S.; Swarnalatha, G.V.; Kavitha, M.D.; Chauhan, V.S.; Ravi, R.; Bansal, A.K.; Singh, R.; Pande, A.; Ravishankar, G.A.; et al. Characterization of fatty acids and hydrocarbons of Chlorophycean microalgae towards their use as biofuel source. *Biomass Bioenerg.* **2015**, *77*, 75–91. [CrossRef]
28. Abou-Shanab, R.A.I.; El-Dalatony, M.M.; EL-Sheekh, M.M.; Ji, M.K.; Salama, E.S.; Kabra, A.N.; Jeon, B.H. Cultivation of a new microalga, Micractinium reisseri, in municipal wastewater for nutrient removal, biomass, lipid, and fatty acid production. *Biotechnol. Bioproc. E* **2014**, *19*, 510–518. [CrossRef]
29. Yoshimura, T.; Okada, S.; Honda, M. Culture of the hydrocarbon producing microalga Botryococcus braunii strain Showa: Optimal CO_2, salinity, temperature, and irradiance conditions. *Bioresour. Technol.* **2013**, *133*, 232–239. [CrossRef]
30. Zhu, L.D.; Hiltunen, E.; Shu, Q.; Zhou, W.Z.; Li, Z.H.; Wang, Z.M. Biodiesel production from algae cultivated in winter with artificial wastewater through pH regulation by acetic acid. *Appl. Energ.* **2014**, *128*, 103–110. [CrossRef]
31. Yang, I.S.; Salama, E.S.; Kim, J.O.; Govindwar, S.P.; Kurade, M.B.; Lee, M.; Roh, H.S.; Jeon, B.H. Cultivation and harvesting of microalgae in photobioreactor for biodiesel production and simultaneous nutrient removal. *Energ. Convers. Manag.* **2016**, *117*, 54–62. [CrossRef]
32. Papadimitriou, C.A.; Papatheodouiou, A.; Takavakoglou, V.; Zdragas, A.; Samaras, P.; Sakellaropoulos, G.P.; Lazaridou, M.; Zalidis, G. Investigation of protozoa as indicators of wastewater treatment efficiency in constructed wetlands. *Desalination* **2010**, *250*, 378–382. [CrossRef]

33. Ji, M.K.; Yun, H.S.; Park, Y.T.; Kabra, A.N.; Oh, I.H.; Choi, J. Mixotrophic cultivation of a microalga Scenedesmus obliquus in municipal wastewater supplemented with food wastewater and flue gas CO2 for biomass production. *J. Environ. Manag.* **2015**, *159*, 115–120. [CrossRef] [PubMed]
34. Wang, L.; Li, Y.C.; Chen, P.; Min, M.; Chen, Y.F.; Zhu, J.; Ruan, R.R. Anaerobic digested dairy manure as a nutrient supplement for cultivation of oil-rich green microalgae *Chlorella* sp. *Bioresour. Technol.* **2010**, *101*, 2623–2628. [CrossRef] [PubMed]
35. Abou-Shanab, R.A.I.; Hwang, J.H.; Cho, Y.; Min, B.; Jeon, B.H. Characterization of microalgal species isolated from fresh water bodies as a potential source for biodiesel production. *Appl. Energ.* **2011**, *88*, 3300–3306. [CrossRef]
36. Laura, G.P. *Algae: Anatomy, Biochemistry, and Biotechnology*; CRC Press: Boca Raton, FL, USA, 2006.
37. Chisti, Y. Biodiesel from microalgae beats bioethanol. *Trends Biotechnol.* **2008**, *26*, 126–131. [CrossRef] [PubMed]
38. Schreiber, C.; Schiedung, H.; Harrison, L.; Briese, C.; Ackermann, B.; Kant, J.; Schrey, S.D.; Hofmann, D.; Singh, D.; Ebenhöh, O.; et al. Evaluating potential of green alga Chlorella vulgaris to accumulate phosphorus and to fertilize nutrient-poor soil substrates for crop plants. *J. Appl. Phycol.* **2018**, *30*, 2827–2836. [CrossRef]
39. Kim, M.K.; Park, J.W.; Park, C.S.; Kim, S.J.; Jeune, K.H.; Chang, M.U.; Acreman, J. Enhanced production of *Scenedesmus* spp. (green microalgae) using a new medium containing fermented swine wastewater. *Bioresour. Technol.* **2007**, *98*, 2220–2228. [CrossRef]
40. Feng, Y.; Li, C.; Zhang, D. Lipid production of *Chlorella vulgaris* cultured in artificial wastewater medium. *Bioresour. Technol.* **2011**, *102*, 101–105. [CrossRef]
41. Singh, P.; Guldhe, A.; Kumari, S.; Rawat, I.; Bux, F. Investigation of combined effect of nitrogen, phosphorus and iron on lipid productivity of microalgae *Ankistrodesmus falcatus* KJ671624 using response surface methodology. *Biochem. Eng. J.* **2015**, *94*, 22–29. [CrossRef]

© 2019 by the authors. Licensee MDPI, Basel, Switzerland. This article is an open access article distributed under the terms and conditions of the Creative Commons Attribution (CC BY) license (http://creativecommons.org/licenses/by/4.0/).

Article

Microalgae Cultivation in Pilot Scale for Biomass Production Using Exhaust Gas from Thermal Power Plants

Sanghyun Park [1,2], Yongtae Ahn [1], Kalimuthu Pandi [1], Min-Kyu Ji [3], Hyun-Shik Yun [4] and Jae-Young Choi [1,*]

1. Center for Environment, Health and Welfare Research, Korea Institute of Science and Technology, Seoul 02792, Korea; shsh1020@kist.re.kr (S.P.); ytahn@kist.re.kr (Y.A.); kpandi@kist.re.kr (K.P.)
2. Graduate School of Energy and Environment (KU-KIST GREEN SCHOOL), Korea University, Seoul 02841, Korea
3. Environmental Assessment Group, Korea Environment Institute, Yeongi-gun 30147, Korea; mkji@kei.re.kr
4. Dongmyung ent. Co., Ltd., Seoul 06254, Korea; awyhs81@gmail.com
* Correspondence: jchoi@kist.re.kr; Tel.: +82-2-958-5846

Received: 29 July 2019; Accepted: 9 September 2019; Published: 11 September 2019

Abstract: Exhaust gases from thermal power plants have the highest amount of carbon dioxide (CO_2), presenting an environmental problem related to a severe impact on ecosystems. Extensively, the reduction of CO_2 from thermal power plants has been considered with the aid of microalgae as a cost-effective, sustainable solution, and efficient biological means for recycling of CO_2. Microalgae can efficiently uptake CO_2 and nutrients resulting in high generation of biomass and which can be processed into different valuable products. In this study, we have taken *Nephroselmis* sp. KGE8, *Acutodesmus obliquus* KGE 17 and *Acutodesmus obliquus* KGE32 microalgae, which are isolated from acid mine drainage and cultivated in a photobiological incubator on a batch scale, and also confirmed that continuous culture was possible on pilot scale for biofuel production. We also evaluated the continuous culture productivity of each cultivate-harvest cycle in the pilot scale. The biomass of the cultivated microalgae was also evaluated for its availability.

Keywords: biomass; microalgae; photobioreactor; power plant exhaust gas; lipid; FAME

1. Introduction

Nowadays, an increase of CO_2 levels in the atmosphere is extensively recognized as a major contributor for global warming. Recent reports are highlighted that atmosphere contains CO_2 level of 450 ppm [1]. The atmospheric CO_2 can be trapped by green plants via photosynthesis. However, terrestrial plants are estimated to reduce only 3–6% of global CO_2 emissions, which is significant given the slow growth rates of plants. On the other hand, microalgae can grow much faster than terrestrial plants, and their CO_2 reduction efficacy was 10–50 times higher than plants [2,3]. The variety of microalgae cultivated in comfortable environmental condition to produce comparably 15–300 times higher energy sources than plants, which also reduce the land area for cultivation and continuously increase the yield per unit area [2,4,5]. Microalgae can biologically store CO_2 through photosynthesis in the form organic compounds and then use microalgal biomass as a feedstock for renewable energy after CO_2 fixation [6]. Moreover, microalgae have been documented as source of valuable biomaterials such as fertilizers, live feed, medicines, and other value-added products.

The large-scale microalgae culture system was divided into two systems, namely the open and closed systems. In the case of the open system, it is difficult to control the amount of light intensity and it may vary depend upon the local time, and also difficult to maintain the temperature. The closed system, which is a device to overcome these limitations, is able to control the light intensity, external influence, and temperature, though the operation cost and the manufacturing cost are high when compared to the open system. Especially, the closed system microalgal growth rate is 1.5–4 times higher than the open system [7]. The high growth rate of microalgae has a large impact on CO_2 capture and may lead to an increase in biomass production. According to the various research condition, closed system may be designed as airlift column, horizontal tube, stirred tank, and flat panel photobioreactor (PBR) [7,8].

Obviously, industrial exhaust gases contain 10–20% of CO_2 with trace amounts of SO_x and NO_x. The selection of microalgae plays a vital role in CO_2 reduction efficacy and represents a significantly cost-effective route for biomass production. The desirable qualities of microalgae comprise high growth and CO_2 consumption rates, also patience towards trace constituents of exhaust flue gases such as SO_x and NO_x and production of valuable products. Maeda et al. (1996) used *Chlorella* sp. T-1 as a potential microalga for the biological removal of exhausted CO_2 from coal-fired thermal power plants. Aslam et al. (2017) have identified that mixed microalgae societies like *Desmodesmus* spp. can slowly grow in 100% unfiltered exhausted gas from coal combustion with phosphate buffering condition [9]. Kassim and Meng (2017) studied biofixation of CO_2 by two microalgae species such as *Chlorella* sp. and *Tetraselmis suecica* with various CO_2 concentration [10]. Even though the above said studies have been carried out in exhausted gas which adversely affects the microalgal growth. To the best of our knowledge, no study has yet reported on the actual injection of exhaust gas, and there is a lack of research on biomass tendency when continuously injecting the gas into large-scale bioreactor.

Hence, the objective of this study is to evaluate the feasibility of microalgae species like *Nephroselmis* sp. KGE8, *Acutodesmus obliquus* KGE 17 and *Acutodesmus obliquus* KGE32, which were cultivated in a laboratory with the supplementation of power plant exhaust gas. Then, evaluate the growth potential of the microalgae in the semi-continuous photobioreactor (PBR) operating with the exhaust gas injection, and evaluate the microalgae productivity at each cultivate-harvest cycle. Finally, we also assessed the feasibility of biodiesel, lipid and C16-18-FAME contents in recovered microalgae.

2. Materials and Methods

2.1. Conditions of Microalgae Cultivation in Batch Scale

Microalgae species were derived from acid mine drainage which include *Nephroselmis* sp. KGE8, *Acutodesmus obliquus* KGE 17, and *Acutodesmus obliquus* KGE32. The batch type cultivation was performed in 140 mL serum bottle with 100 mL of Bold's Basal Medium (BBM) [11] which contained strains with optical density of 0.010 in UV spectroscopy 680 nm region.

The exhaust gas from the Y coal-fired thermal power plant (Gangwon-do, South Korea) was used as a carbon source for microalgae growth. The collected gas was filtered through a 0.2 μm filter and then supplied at a flow rate of 0.5 L min^{-1} for 1 h to complete saturation of the medium. Composition of the exhaust gas details appeared in Table 1. The microalgae cultivation was conducted at 25 °C with 120 μmol photon m^{-2}s^{-1} of light intensity, and the content was agitated in incubator shaker (Witeg, Wisecube WIS-ML, Germany) at 120 rpm to prevent agglomeration for 7 days.

Table 1. Y Power plant gas contents and concentrations located in Gangwon-do.

Gas Contents	Initial Gas Concentration
CO_2 (v/v %)	14.9 (±0.2)
NO_x (ppmv)	220.2 (±10.5)
SO_x (ppmv)	32 (±5)
CO (ppmv)	1549 (±242)
O_2 (%)	5.47 (±0.03)

2.2. Cultivation of Microalgae in Photobioreactor (PBR)

The culture system used in this study was a multistage photobioreactor (PBR), each PBR has capacity of 2000 L and it is shown in Figure 1. Initially, exhausted gas was injected through PBR 1 and then subsequently passed through the PBR 2, -3 and -4 respectively. Finally, unutilized gas was ejected from the PBR 5, the concentration of CO_2 in each stage was measured and attached in Table 2. The initial and final exhaust gas concentration was analyzed by Testo 350 K emission analyzer (Testo, Germany). According to Blair et al. (2014) red light emitting diode (LED) was installed at each stage for maximize light absorption [12]. The first stage of the pilot scale, the growth rate of each stage of *Nephroselmis* sp. KGE 8 was determined and grown for 20 days. The second cultivation was performed for 22 days, from the 18 days when the specific growth rate (μ_{max}) started to increase, the possibility of continuous cultivation was evaluated by harvest and regrowth. On day 18, the microalgae were recovered and diluted and re-cultured for 16 days and supplemented with BBM to prevent nutrition loss. The interval between collection and incubation was 2 days, and cultures were collected and cultivated three times.

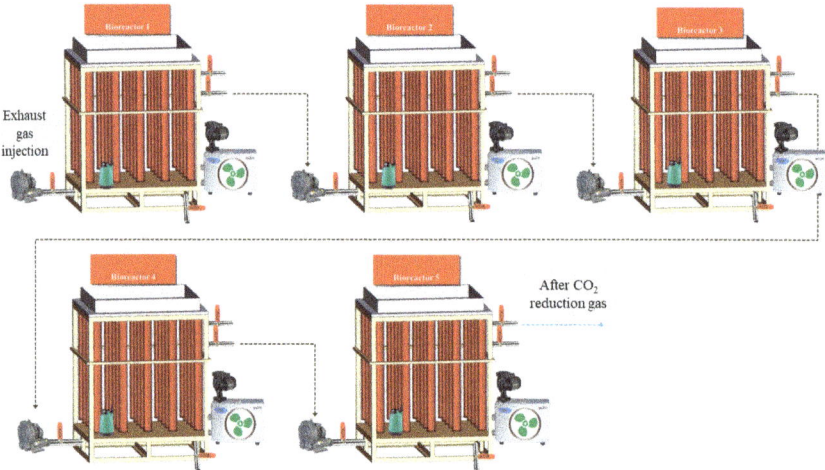

Figure 1. Pilot scale multi-step reactor schematic.

Table 2. Injection CO_2 concentration in each stage.

Stage	CO_2 (v/v %)	NO_x (ppmv)	CO (ppmv)	O_2 (v/v %)
PBR1	14.90 (±0.18)	220.2 (±10.5)	1548.5 (±242)	5.47 (±0.03)
PBR2	8.08 (±0.52)	124.4 (±21.8)	941.1 (±41.9)	12.42 (±0.93)
PBR3	6.72 (±0.73)	99.3 (±16.0)	703.3 (±51.4))	13.89 (±0.99)
PBR4	4.99 (±0.52)	72.3 (±12.1)	537.2 (±21.6)	15.81 (±0.86)
PBR5	3.87 (±0.52)	51.9 (±9.3)	421.1 (±35.2)	17.03 (±0.95)
Out	3.11 (±0.60)	51.8 (±8.18)	337.0 (±5.1)	16.86 (±1.41)

2.3. Analysis of Microalgal Growth

The growth rate of microalgae cultivated in pilot plant was obtained by analysis of OD_{680}, from spectrophotometer (Hach DR/2800, Loveland, CO, USA) which values were converted to dry cell weight (DCW) concentration (g L^{-1}). DCW of *Nephroselmis* sp. KGE 8 was calculated by:

$$\text{Dry weight (g L}^{-1}) = 0.3997 \times OD_{680} - 0.0471 \ (R^2 = 0.9871) \tag{1}$$

Further, the specific growth rate (SGR) was calculated by Equation (2):

$$\mu = (\ln X_2 - \ln X_1)/(t_2 - t_1) \tag{2}$$

where X_1 and X_2 are the mass of initial and final weight of microalgae, respectively, which is used to calculate DCW in this study, and t_1 and t_2 are the initial and final incubation times respectively.

2.4. Algal Harvest

In pilot scale, the algal harvest was performed by sludge pump, and collected to storage tank. The algae of each stage were harvested, and the harvested algae were precipitated and recovered by separating the supernatant and algae.

2.5. Analysis of Lipid and C16–C18 Fatty Acid Methyl Ester (FAME)

The modified Bilgh and Dyer method (Ji et al. (2016)) was used to analyze the lipids and fatty acids in the harvested microalgae [13]. Fatty acids were identified by the modified Lepage and Roy method from Yun et al. (2015), which convert fatty acid into fatty acid methyl esters through esterification and is analyzed by Gas chromatography with a flame ionization detector (GC-FID) using HP-INNOWax capillary column (Agilent Technologies, USA) [14].

3. Results and Discussion

3.1. Growable Microalgae in Exhaust Gas Condition

Batch scale experiment results shows that both *Nephroselmis* sp. KGE 8 and *Acutodesmus obliquus* KGE 17 have lag phase up to two days and showed exponential growth phase until fifth day (Figure 2). At this moment, *Nephroselmis* sp. KGE 8 exhibited the maximum growth when compared to the *Acutodesmus obliquus* KGE 17. Another microalga like *Acutodesmus obliquus* KGE 32 exhibits the lag phase until three days, and the exponential growth phase was until five days, and it has the stationary growth phase. The growth rates of microalgae with a supply of exhaust are presented in Table 3. Ji et al. (2017) and Yun et al. (2016) evaluated the potential for biofuel production according to changes of CO_2 concentration in exhaust gas. Compare with these previous studies, algae production was faster. Also Tang et al. (2011) was focused on growth potential in high concentration of CO_2 and effective concentration of CO_2, growth rate and lipid contents was lower than this study. *Nephroselmis* sp. KGE 8 have the OD_{680} value of 1.341 and the maximum specific growth rate (μ_{max}) was 1.41 d^{-1} between 3–4 days of culture. Conversely, *Acutodesmus obliquus* KGE 32 and *Acutodesmus obliquus* KGE 17 possess the OD_{680} values of 0.970 and 0.553 and μ_{max} were 1.08 d^{-1} and 1.37 d^{-1} respectively. The microalgae, which applied in this study, showed higher specific grow rates (1.08 to 1.37 d^{-1}) than previous study (Table 3). Continuous and excessive exposure of NO_x and SO_x gases to cells could leads to inhibition of microalgae growth rate [15,16]. Praveenkumar et al. (2014a) reported that algal FAME content and productivity increased from 129 to 168 mg fame/g cells and from 59 to 118 mg fame/L d, respectively, in coal-fired flue-gas inlet condition [17]. They also conclude that stress conditions could lead to improve algal lipid productivity [18].

3.2. Pilot Scale Cultivation

The pilot scale cultivation result discloses that, lag phase period of *Nephroselmis* sp. KGE 8 was increased from 2 days to 10 days when compared with batch scale results due to the stress present in the exhaust gas (Figure 3). Same trend was also observed in previous study by Borowitzka et al. (2018) and mentioned that adaptation by stress due to CO_2 [21].

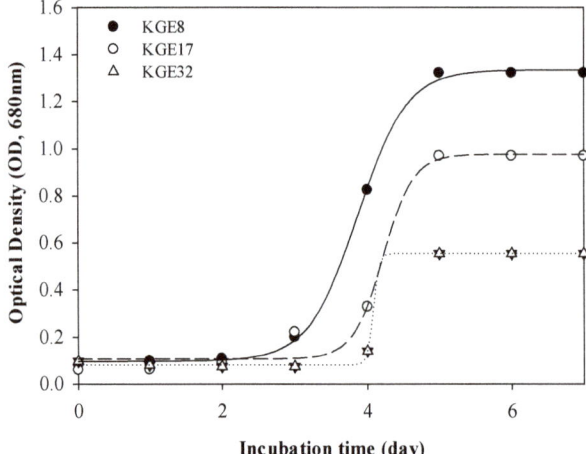

Figure 2. Growth curve of *Nephroselmis* sp. KGE 8, *Acutodesmus obliquus* KGE 17, and *Acutodesmus obliquus* KGE 30 microalgae in purged power plant gas.

Table 3. Specific growth rate and lipid contents of various microalgal strains cultivated at different carbon dioxide concentration.

Species	Carbon Dioxide Concentration (%)	Incubation Condition	Medium	Lipid Contents (%)	Specific Growth Rate (μ_{max}, d^{-1})
Scenedesmus obliquus KGE 9 [a]	14.1	Batch	BBM	22.8	1.00
Chlorella pyrenoidosa SJTU-2 [b]	10.0	Batch	BG11	24.2	0.78
Acutodesmus obliquus KGE 30 [c]	14.1	Batch	BBM	17.5	1.09
Acutodesmus obliquus KGE 32 [d]	14.1	Batch	BBM	-	1.08
Acutodesmus obliquus KGE 17 [d]	14.1	Batch	BBM	-	1.37
Nephroselmis sp. KGE 8 [d]	14.1	Batch	BBM	59.4	1.41
0 *Nephroselmis* sp. KGE 8 [d]	14.1	Pilot scale	BBM	60.9	0.26

[a] Ji et al. (2017) [19]; [b] D. Tang et al. (2011) [20]; [c] HS Yun et al. (2016) [14]; [d] This study.

Further, the microalgae growth was not similar with the batch scale (Figure 3). Due to the different character of coal and also the generated exhaust gas from the thermal power plant does not contain constant amount of CO_2, it may lead to irregular growth of microalgae in the pilot scale. Cheng et al. (2019) also reported that biomass yields were not constant for every cycle, even gas-adapted microalgae were injected with a constant concentration of mixed gas [22]. These results indicated that *Nephroselmis* KGE 8 is a microalga species that could adaptively grow, even when the exhaust gas was continuously injected.

In continuous culture potential evaluation experiment, *Nephroselmis* sp. KGE 8 reached the exponential growth phase at 17 days after initiated the cultivation (Figure 4). According to Tan et al. (2018), the productivity of microalgae tended to decrease with increasing amount of cultivation [23]. The growth of *Nephroselmis* sp. KGE 8 was different in each stage. PBR 2 show a microalgae concentration of 0.6002 g L^{-1} for the first time and 0.4932 g L^{-1} for the second cultivation. Also the microalgae concentration in PBR 3, PBR 4, and PBR 5 was decreased from 0.5644 g L^{-1} to 0.4955g L^{-1}, 0.5343 g L^{-1} to 0.4722 g L^{-1}, and 0.4421 g L^{-1} to 0.4116 g L^{-1} respectively. In contrast, the microalgae growth in PBR 1 has increased from 0.4996 g L^{-1} to 0.5710 g L^{-1}, unlike other stages. Biomass productivity is affected by growth factors, and according to Sun et al. (2018), the growth factors like high temperature and large N source will increase biomass growth [24]. When NO_x is dissolved in water, it tends to form nitrite, which can help to grow the biomass [25].

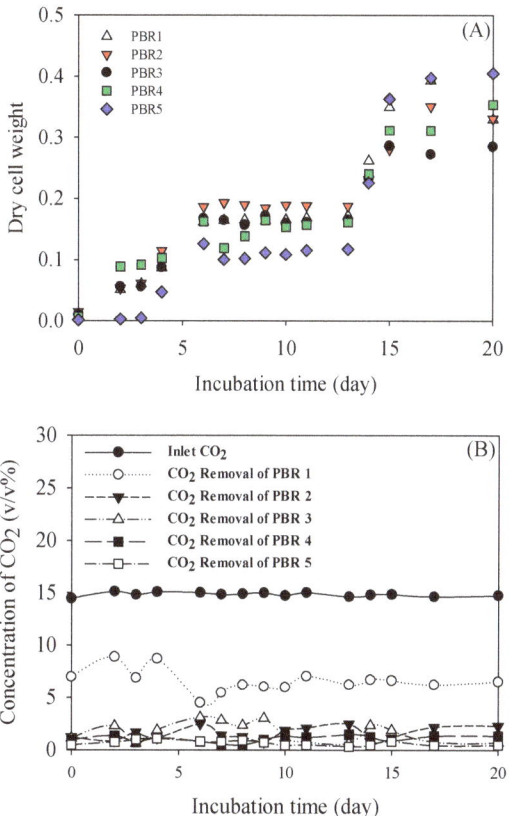

Figure 3. The growth of *Neproselmis* sp. KGE 8 when injecting exhaust gas from thermal power plant using multi-step reactor. Growth rate was shown to (**A**), and removal CO_2 concentration was shown to (**B**).

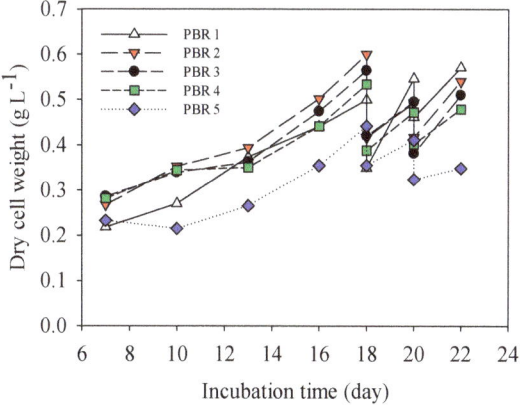

Figure 4. Microalgae growth curve in each incubator re-cultivation process.

3.3. Lipid and Fatty Acid Productivity

We have compared the biomass productivity of *Nephroselmis* sp. KGE 8 in both batch and pilot scale. In batch scale, the biomass productivity was found to be 0.696 g L^{-1} d^{-1}, also contain 59.4% and 95.1% of lipid and C16–C18 FAME respectively. In contrast, the biomass productivity, lipid, and C16-C18 FAME contents were decreased to 0.163 g L^{-1} d^{-1}, 39.4%, and 77.8% respectively in pilot scale experiments (Table 4). Park et al. (2013) reported that the maximum lipid content of the *Nephroselmis* sp. KGE 8 species was 38.8% [26].

Table 4. Biomass productivity, C16-C18 ratio, and lipid content of *Nephroselmis* sp. KGE8 at laboratory scale and pilot scale.

Strain	Volumetric Productivity of Biomass at μ_{max} (g L^{-1} day^{-1})	C16–C18 Ratio (wt %)	Lipid Content (wt %)
KGE8 cultivated in Laboratory	0.696	95.1	59.4
KGE8 Cultivated in Pilot scale	0.163	77.8	60.9

The average lipid content of harvested microalgae in PBR 1 was 41.24%, which was higher than that of other stages. Arief et al. (2009) reported that the content of lipid in microalgae increased with increasing CO_2 concentration. The fatty acid content of the recovered *Nephroselmis* sp. KGE 8 illustrated in Figure 5. The average of C16 to C18 FAME contents in recovered microalgae at each harvest cycle illustrated Figure 5A. The fatty acid content was 74.38% (*w*/*w*), and the highest fatty acid content showed the highest fatty acid content as 87.29% in PBR 5. However, as the number of continuous cultures increased, the fatty acid content of PBR 1 also increased, while that of PBR 2 to PBR 5 tended to decrease (Figure 5B). In particular, PBR 5, which showed the highest fatty acid content at the initial stage, showed a sharp decrease in fatty acid content as the number of times increased. Sharmaet al. (2012), and Nayaket al. (2018) reported an increase in Oleic acid (C18: 1) in the fatty acids of cultured algae in coal combustion gases [27,28].

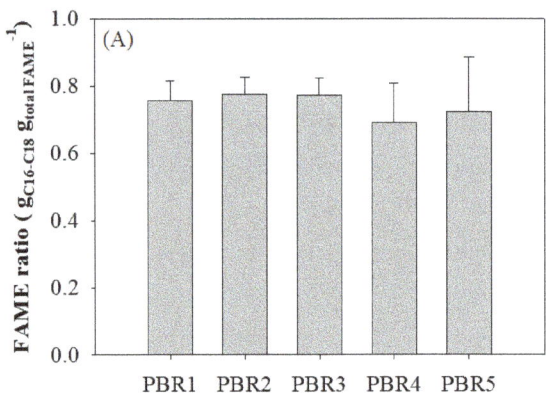

Figure 5. *Cont.*

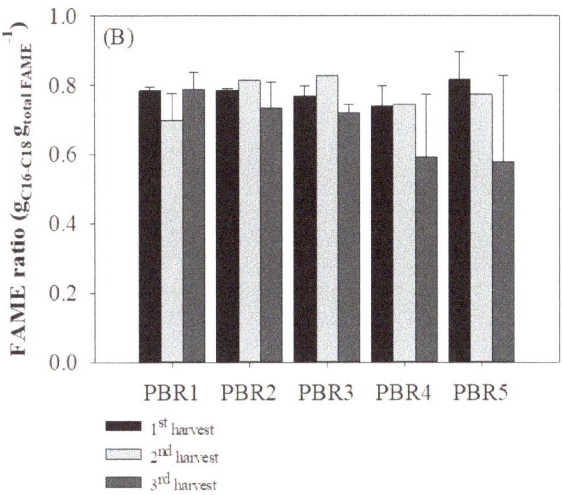

Figure 5. C16–C18 FAME yield from each photo bioreactor.

4. Conclusions

Batch scale studies reveal that *Nephroselmis* sp. KGE 8 showed the best growth under exhaust gas conditions. *Nephroselmis* sp. KGE 8 showed growth potential (0.696 g L^{-1}) in the semi-continuous PBR operation with the exhaust gas injection. The lipid content and C16-C18 FAME content were 39.4% and 77.8% in PBR1, respectively. The microalgae productivity of five reactors showed range from 0.4116 g L^{-1} to 0.5468 g L^{-1} at each cultivate-harvest cycles. PBR 1 showed highest microalgae productivity during PBR operation.

When exhaust gas is directly injected, changes in NO$_x$ and temperature condition accelerate the microbial energy conversion. Singh et al. (2014) reported that some algal species obtained maximum biomass in 15% CO$_2$ [29]. Based on the result, it was concluded that direct injection of exhaust gas is the most suitable condition for utilization of energy source of microalgae. This microalgal cultivation system could be a suitable process for the massive cultivation of microalgae with exhaust gas from power plants.

Author Contributions: For research articles with several authors, a short paragraph specifying their individual contributions must be provided. The following statements should be used "Conceptualization, S.P. and H.-S.Y.; Methodology, H.-S.Y.; Software, S.P.; Validation, H.-S.Y.; Investigation, M.-K.J.; Data Curation, S.P.; Writing-Original Draft Preparation, S.P.; Writing-Review & Editing, Y.A. and K.P.; Visualization, H.-S.Y.; Supervision, J.-Y.C.; Project Administration, J.-Y.C.; Funding Acquisition, J.-Y.C.".

Funding: This work was supported by the Korea CCS R&D Center (Korea CCS 2020 Project) grant funded by the Korea government (Ministry of Science, ICT & Future Planning) in 2019 (KCRC-2014M1A8A1049293) and the KIST Institutional Program (Project No. 2E29670).

Conflicts of Interest: The authors declare no conflict of interest.

References

1. Shen, Y. Carbon dioxide bio-fixation and wastewater treatment via algae photochemical synthesis for biofuels production. *RSC Adv.* **2014**, *4*, 49672–49722. [CrossRef]
2. Yuan, C.; Wang, S.; Cao, B.; Hu, Y.; Abomohra, A.E.-F.; Wang, Q.; Qian, L.; Liu, L.; Liu, X.; He, Z.; et al. Optimization of hydrothermal co-liquefaction of seaweeds with lignocellulosic biomass: Merging 2nd and 3rd generation feedstocks for enhanced bio-oil production. *Energy* **2019**, *173*, 413–422. [CrossRef]

3. Cuellar-Bermudez, S.P.; Garcia-Perez, J.S.; Rittmann, B.E.; Parra-Saldívar, R. Photosynthetic bioenergy utilizing CO_2: An approach on flue gases utilization for third generation biofuels. *J. Clean. Prod.* **2015**, *98*, 53–65. [CrossRef]
4. Chisti, Y. Biodiesel from microalgae. *Biotechnol. Adv.* **2007**, *25*, 294–306. [CrossRef]
5. Yun, H. Carbon Dioxide Sequestration of Exhaust Fumes by Freshwater Microalgae with Massive Biomass Production. Bachelor's Thesis, Yonsei University, Seoul, Korea, 2015.
6. Zhao, B.; Su, Y.; Zhang, Y.; Cui, G. Carbon dioxide fixation and biomass production from combustion flue gas using energy microalgae. *Energy* **2015**, *89*, 347–357. [CrossRef]
7. Chew, K.W.; Chia, S.R.; Show, P.L.; Yap, Y.J.; Ling, T.C.; Chang, J.-S. Effects of water culture medium, cultivation systems and growth modes for microalgae cultivation: A review. *J. Taiwan Inst. Chem. Eng.* **2018**, *91*, 332–344. [CrossRef]
8. Zhao, B.; Su, Y. Process effect of microalgal-carbon dioxide fixation and biomass production: A review. *Renew. Sustain. Energy Rev.* **2014**, *31*, 121–132. [CrossRef]
9. Aslam, A.; Thomas-Hall, S.R.; Mughal, T.A.; Schenk, P.M. Selection and adaptation of microalgae to growth in 100% unfiltered coal-fired flue gas. *Bioresour. Technol.* **2017**, *233*, 271–283. [CrossRef]
10. Kassim, M.A.; Meng, T.K. Carbon dioxide (CO_2) biofixation by microalgae and its potential for biorefinery and biofuel production. *Sci. Total Environ.* **2017**, *584*, 1121–1129. [CrossRef]
11. Bischoff, H.W. Phycological studies. IV. Some Algae from Enchanted Rock and Related Algal Species. *Univ. Texas Publ.* **1963**, *6318*, 95.
12. Blair, M.F.; Kokabian, B.; Gude, V.G. Light and growth medium effect on Chlorella vulgaris biomass production. *J. Environ. Chem. Eng.* **2014**, *2*, 665–674. [CrossRef]
13. Ji, M.-K.; Yun, H.-S.; Hwang, B.S.; Kabra, A.N.; Jeon, B.-H.; Choi, J. Mixotrophic cultivation of Nephroselmis sp. using industrial wastewater for enhanced microalgal biomass production. *Ecol. Eng.* **2016**, *95*, 527–533. [CrossRef]
14. Yun, H.S.; Ji, M.K.; Park, Y.T.; Salama el, S.; Choi, J. Microalga, Acutodesmus obliquus KGE 30 as a potential candidate for CO_2 mitigation and biodiesel production. *Environ. Sci. Pollut. Res. Int.* **2016**, *23*, 17831–17839. [CrossRef] [PubMed]
15. Li, Y.; Horsman, M.; Wang, B.; Wu, N.; Lan, C.Q. Effects of nitrogen sources on cell growth and lipid accumulation of green alga Neochloris oleoabundans. *Appl. Microbiol. Biotechnol.* **2008**, *81*, 629–636. [CrossRef] [PubMed]
16. Chiu, S.-Y.; Kao, C.-Y.; Huang, T.-T.; Lin, C.-J.; Ong, S.-C.; Chen, C.-D.; Chang, J.-S.; Lin, C.-S. Microalgal biomass production and on-site bioremediation of carbon dioxide, nitrogen oxide and sulfur dioxide from flue gas using Chlorella sp. cultures. *Bioresour. Technol.* **2011**, *102*, 9135–9142. [CrossRef] [PubMed]
17. Praveenkumar, R.; Kim, B.; Choi, E.; Lee, K.; Cho, S.; Hyun, J.-S.; Park, J.-Y.; Lee, Y.-C.; Lee, H.U.; Lee, J.-S.; et al. Mixotrophic cultivation of oleaginous Chlorella sp. KR-1 mediated by actual coal-fired flue gas for biodiesel production. *Bioprocess Biosyst. Eng.* **2014**, *37*, 2083–2094. [CrossRef] [PubMed]
18. Praveenkumar, R.; Kim, B.; Choi, E.; Lee, K.; Park, J.-Y.; Lee, J.-S.; Lee, Y.-C.; Oh, Y.-K. Improved biomass and lipid production in a mixotrophic culture of Chlorella sp. KR-1 with addition of coal-fired flue-gas. *Bioresour. Technol.* **2014**, *171*, 500–505. [CrossRef] [PubMed]
19. Ji, M.K.; Yun, H.S.; Hwang, J.H.; Salama, E.S.; Jeon, B.H.; Choi, J. Effect of flue gas CO_2 on the growth, carbohydrate and fatty acid composition of a green microalga Scenedesmus obliquus for biofuel production. *Environ. Technol.* **2017**, *38*, 2085–2092. [CrossRef]
20. Tang, D.; Han, W.; Li, P.; Miao, X.; Zhong, J.-J. CO_2 biofixation and fatty acid composition of Scenedesmus obliquus and Chlorella pyrenoidosa in response to different CO_2 levels. *Bioresour. Technol.* **2011**, *102*, 3071–3076. [CrossRef]
21. Borowitzka, M.A. The 'stress' concept in microalgal biology—Homeostasis, acclimation and adaptation. *Environ. Boil. Fishes* **2018**, *30*, 2815–2825. [CrossRef]
22. Cheng, D.; Li, X.; Yuan, Y.; Yang, C.; Tang, T.; Zhao, Q.; Sun, Y. Adaptive evolution and carbon dioxide fixation of Chlorella sp. in simulated flue gas. *Sci. Total Environ.* **2019**, *650*, 2931–2938. [CrossRef] [PubMed]
23. Tan, X.B.; Lam, M.K.; Uemura, Y.; Lim, J.W.; Wong, C.Y.; Ramli, A.; Kiew, P.L.; Lee, K.T. Semi-continuous cultivation of Chlorella vulgaris using chicken compost as nutrients source: Growth optimization study and fatty acid composition analysis. *Energy Convers. Manag.* **2018**, *164*, 363–373. [CrossRef]

24. Sun, X.-M.; Ren, L.-J.; Zhao, Q.-Y.; Ji, X.-J.; Huang, H. Microalgae for the production of lipid and carotenoids: A review with focus on stress regulation and adaptation. *Biotechnol. Biofuels* **2018**, *11*, 272. [CrossRef] [PubMed]
25. Brown, L.M. Uptake of carbon dioxide from flue gas by microalgae. *Energy Convers. Manag.* **1996**, *37*, 1363–1367. [CrossRef]
26. Park, Y.-T.; Lee, H.; Yun, H.-S.; Song, K.-G.; Yeom, S.-H.; Choi, J. Removal of metal from acid mine drainage using a hybrid system including a pipes inserted microalgae reactor. *Bioresour. Technol.* **2013**, *150*, 242–248. [CrossRef] [PubMed]
27. Sharma, K.K.; Schuhmann, H.; Schenk, P.M. High Lipid Induction in Microalgae for Biodiesel Production. *Energies* **2012**, *5*, 1532–1553. [CrossRef]
28. Nayak, M.; Dhanarajan, G.; Dineshkumar, R.; Sen, R. Artificial intelligence driven process optimization for cleaner production of biomass with co-valorization of wastewater and flue gas in an algal biorefinery. *J. Clean. Prod.* **2018**, *201*, 1092–1100. [CrossRef]
29. Singh, S.; Singh, P. Effect of CO_2 concentration on algal growth: A review. *Renew. Sustain. Energy Rev.* **2014**, *38*, 172–179. [CrossRef]

© 2019 by the authors. Licensee MDPI, Basel, Switzerland. This article is an open access article distributed under the terms and conditions of the Creative Commons Attribution (CC BY) license (http://creativecommons.org/licenses/by/4.0/).

Article

Towards the Development of Syngas/Biomethane Electrolytic Production, Using Liquefied Biomass and Heterogeneous Catalyst

Ana Gonçalves [1], Jaime Filipe Puna [1,2,*], Luís Guerra [3], José Campos Rodrigues [3], João Fernando Gomes [1,2], Maria Teresa Santos [1] and Diogo Alves [1]

- [1] Área Departamental de Engenharia Química, Instituto Superior de Engenharia de Lisboa, Instituto Politécnico de Lisboa, R, Conselheiro Emídio Navarro, 1, 1959-007 Lisboa, Portugal; ana.m.goncalves93@gmail.com (A.G.); jgomes@deq.isel.ipl.pt (J.F.G.); tsantos@deq.isel.ipl.pt (M.T.S.); Diogo_MPAlves@hotmail.com (D.A.)
- [2] CERENA—Centro de Recursos Naturais e Ambiente, Instituto Superior Técnico, Universidade de Lisboa, Av. Rovisco Pais, 1, 1049-001 Lisboa, Portugal
- [3] GSyF, Pol. Ind. Alto do Ameal, Pavilhão C-13, 2565-641 Torres Vedras, Portugal; lguerra@live.com.pt (L.G.); jjcr@outlook.pt (J.C.R.)
- * Correspondence: jpuna@deq.isel.ipl.pt; Tel.: +351-218317254

Received: 27 August 2019; Accepted: 1 October 2019; Published: 6 October 2019

Abstract: This paper presents results on the research currently being carried out with the objective of developing new electrochemistry-based processes to produce renewable synthetic fuels from liquefied biomass. In the current research line, the gas mixtures obtained from the typical electrolysis are not separated into their components but rather are introduced into a reactor together with liquefied biomass, at atmospheric pressure and different temperatures, under acidified zeolite Y catalyst, to obtain synthesis gas. This gaseous mixture has several applications, like the production of synthetic 2nd generation biofuel (e. g., biomethane, biomethanol, bio-dimethyl ether, formic acid, etc.). The behaviour of operational parameters such as biomass content, temperature and the use of different amounts of acidified zeolite HY catalyst were investigated. In the performed tests, it was found that, in addition to the synthesis gas (hydrogen, oxygen, carbon monoxide and carbon dioxide), methane was also obtained. Therefore, this research is quite promising, and the most favourable results were obtained by carrying out the biomass test at 300 °C, together with 4% of acidified zeolite Y catalyst, which gives a methane volumetric concentration equal to 35%.

Keywords: liquefied biomass; electrolysis; synthesis gas; renewable energy; synthetic fuels; HY zeolite

1. Introduction

Oil-derived fuels are essential for complying with the World's energy needs, accounting for the majority (more than 80%) of the global primary energy consumption, and recent forecast studies, developed by the IEA (2017) [1] and BP (2018) [2], show a continuing growth in fossil fuel demand [3], in the near future, considering a wide range of factors such as demand, technology development, assumptions of policy agreements in order to reduce greenhouse gas emissions (GHG), as well, as changes in the regional production capacity [4]. In spite of this dependence, recent concerns over climate change have driven society to seek for alternatives in order to reduce GHG emissions. This resulted in a continuous search for a shift in energy production from fossil fuels toward renewables [5].

Consequently, the use of biomass as a source of renewable energy has recently been increasing. When compared to fossil fuels, biomass energy has several advantages which includes its renewable nature, carbon neutral ability, low sulfur emission during combustion, relative abundance and its easy transportation and storage. Therefore, biomasses are, potentially, one of the more important available resources to produce new liquid biofuels, synthesis gas (syngas), biohydrogen, solid biofuels, and, valuable chemicals [6]. Liquefaction is a relatively novel process capable of converting biomass into bio-oil products [7]. In general, the liquefaction of biomass consists of three main steps: depolymerization followed by decomposition, and, recombination at high temperatures [8]. Normally, the biomass liquefaction processes use a specific solvent, such as water or organic ones, such as, methanol, ethanol, phenol, acetone, etc., to interact strongly with the biomass components [9]. More recently, cork by-products, have been reported as interesting raw materials for liquefaction, by conventional, microwave induced, as well as ultrasounds-assisted methods [6]. The main components of these solid biomass, like cork and eucalyptus bark are, lignin, cellulose and hemi-cellulose [9].

The bio-oil obtained from the liquefaction processes of these solid biomass components, performed at 160 °C and 90 min, such as described by Mateus et al. [6], has several advantages, like its utilization as fuel, in engines and, in other combustion units, such as, boilers, furnaces, etc., as auxiliary fuel or, can be converted into high quality chemical products, through several processes, like, catalytic cracking, hydrogenation or steam reforming [10].

Regarding electrolysis process, the main four technologies developed are, alkaline electrolysis, Proton-exchange membrane (PEM), Solid Oxide Electrolysis (SOE) [11,12] and, finally, Polymeric Anion Exchange Membrane (AEM) processes [13].

The syngas applications are several, not only related with synthetic biofuels production, but also, in the added-value chemical products, such as, formic acid, ethylene, but also, methyl acetate, acetic acid, formaldehyde and polyolefins (these last four produced from biomethanol) [14]. In the field of synthetic biofuels production, syngas can be converted into biomethanol [15,16], bio-DME [17], biomethane (synthetic natural gas) through the Sabatier process [18,19], but also, into biodiesel, bio-gasoline, bio-naphtha, etc., through the Fischer-Tropsch process [20,21].

This paper describes a further new approach on a new technology, previously reported by the authors [22] capable of producing syngas in a single step, without separation of the elementary gases, produced during the water alkaline electrolysis. It is called co-electrolysis of water, under the alkaline process, using a carbon source to directly produce the syngas mixture, at low temperatures and pressures, thus requiring significatively less amounts of energy inputs [11,23]. This previous approach, uses graphite electrodes, as a source of carbon, that is further oxidized, during the electrolysis process, to carbon monoxide and carbon dioxide which are present in the generated gas mixture (syngas), and, efficiently converts electricity from renewable sources (mainly wind or solar, or when this electricity is in excess in the electrical grid, or in off peak hours). Thus, this new technology is able to convert electricity into syngas, which is an intermediate for the generation of synthetic 2nd G biofuels, which was already demonstrated [24]. The main drawback is the (small) consumption of the graphite electrodes and its relatively high cost, which could be avoided if steel electrodes are used together with an additional carbon source, such as, liquefied biomass, to be added in the electrolyser. Concerning the use of liquefied biomass, some results from preliminary trials have been recently published elsewhere [25], and points out that, the process needs enhancement, such as, the use of solid catalysts. In this new process, the gas obtained from electrolysis is not separated into its components and, it's introduced into a reactor together with a specific content of a previous mixture of cork/eucalyptus bark liquefied biomass, at normal pressure and different temperatures. The gas is released upon contact with the biomass, thus resulting into syngas, which is a mixture consisting essentially of carbon monoxide, hydrogen, carbon dioxide and some unreacted oxygen. In this work, the behaviour of operational parameters such as biomass content and type, temperature and the use of different amounts of acidified zeolite (z.) HY catalyst were investigated. In the performed tests, it was found that, in addition to the syngas, methane was also produced, with significant content. The purpose of use samples of different kinds of liquified biomass (described in Section 2.1) of cork and/or eucalyptus bark, with and without the correspondent sugars solubilized in aqueous solvent, was to investigate if, there was significant influence in the output syngas/methane produced, at the methanation reactor, as well, the influence of temperature and catalyst content in this process. The temperatures range choose for this study must be significantly lower than the typical temperatures used in the gasification process (700–800 °C). The advantage of this technology is located, precisely, in the utilization of lower/medium temperatures, when compared with the coal/biomass gasification and steam reforming processes, which produces, also, syngas. The utilization of lower temperatures will lead to significant input energy savings to the process and, as consequence, lower operating costs. On the other hand, the influence of using lower catalyst contents in the methanation process in this study, is to see if the methane concentration will increase in these temperatures, with and without catalyst.

2. Materials and Methods

2.1. Chemicals

The chemicals used in this research work were, sodium hydroxide (pellets) from VWR Chemicals Prolabo (Fontenay-sous-Bois, France), Y powdered zeolite in the basic form (NaY), from Sigma-Aldrich (Darmstad, Germany) and ammonium nitrate from Merck (Darmstad, Germany). The biomass liquified samples employed were four, one was a mixture of cork and eucalyptus bark with the correspondent sugars (A1), other obtained from the liquefaction process of only cork biomass, with the correspondent sugars (A2), other obtained from the liquefaction process of only eucalyptus bark, without the correspondent sugars and solvents previously removed (A3), and, finally, the last one, similar to A3 but with the correspondent sugars and a significant quantity of solvent (A4). These liquified biomass samples were obtained through a hydro-liquefaction process, with an organic solvent and, also, with an acid homogeneous catalyst, in a range of temperatures between 160–200 °C.

2.2. Syngas and Methane Production Equipment

The production of syngas was carried out on a laboratory apparatus, schematically shown in Figure 1, consisting of: (i) a cylindrical storage tank feeding electrolyte to the solution; (ii) a second storage for the electrolyte solution; (iii) a column containing molecular sieve, in order to adsorb the humidity of the produced gas, and finally, (iv) an electrolyser, where the various electrochemical reactions take place to produce synthesis gas. The electrolyser has a total of seven steel electrodes, each one with a diameter of 5 cm and, a thickness of 0.2 cm, thus resulting an area of 20 cm^2 by electrode, forming disks with two holes each, thus creating electrolyte circulation channels. One channel is connected to the electrolyser input, while the other is connected to its output, thus allowing the out flow of produced gases. The electrodes are spaced from each other 0.3 cm, thus creating in the electrolyser, 8 electrolytic cells. The body of the vessel is made of methyl methacrylate polymer, with a basis of stainless steel, to withstand pressure. To prevent heat losses from the electrolyser and the electrolyte circulation tank, the components are insulated with rockwool. The methanation reactor (v) used is made of glass and consists of an inlet pipe that extends into the reactor where the gas will bubble and an outlet pipe for the produced gas (syngas) that goes to a condenser (vi) with a coupled tank where the resulting condensation and the final gas for analysis are collected. This reactor has approximately 7.6 cm of diameter and 5.7 cm of high (Figure 2). During operation, the electrolyte is admitted through the inlet valves, thus filling up the electrolyser. Then, electric terminals connect the electrodes to the power supply source and the electrolysis process takes place. The produced gas composition is measured by specific sensors (CLEVER CY-12C oxygen analyser from CLEVER, Beijing, China, carbon dioxide and carbon monoxide analysers, both from KELISAIKE (Beijing, China), and a methane analyser from *Exibd R* (Beijing, China), previously calibrated and validated by gas chromatography and, the total flow rate of gases were measured by a volume displacement device. In this study, liquefied biomass from cork wastes was used, obtained as described elsewhere [6]. In order to improve the composition of the produced syngas, a solid catalyst was used, which was prepared from a powdered Y zeolite in basic form, that was acidified using ammonium nitrate, by a traditional technique [26]. Figure 3, shows, on the left, the electrolyser unit and, on the right, the used steel electrodes prepared, like described above.

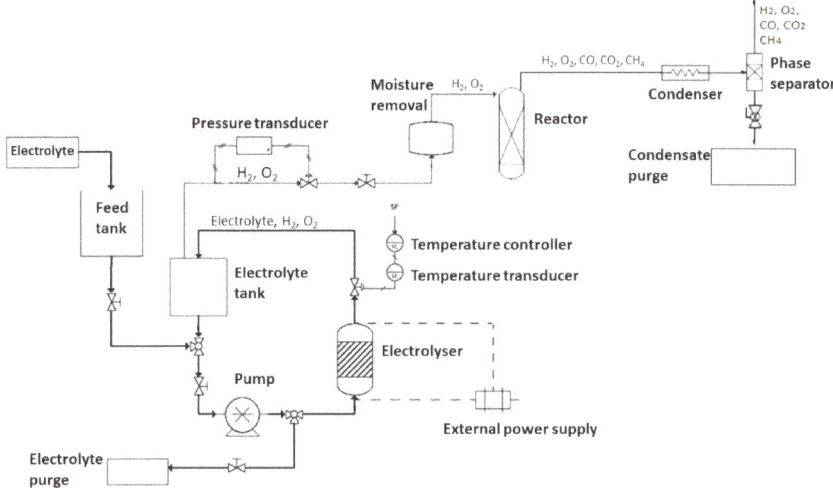

Figure 1. Experimental set-up.

Figure 2. Glass reactor used with liquified biomass to produce methane.

Figure 3. Electrolyser (left) and, steel electrodes used in the electrolyser unit (right).

The measures of pH and conductivity of the electrolyte solutions in the electrolyser in the beginning and the end of each experiment, were conducted, respectively, with a HANNA Instruments (Woonsocket, RI, USA) portable device and, with a GLP32 conductimeter (Crison. Barcelona, Spain). A PB 3002 balance (Mettler Toledo, Columbus, OH, USA) with a precision of 0.01 mg was also used to weigh the solid samples and, also, an oven and a furnace, from Nabertherm (Lilienthal, Germany), were used to dry and calcinate, respectively, the solid catalytic samples.

2.3. Solid and Liquid Samples Characterization

Regarding catalyst characterisation, zeolite Y acidified was characterised by Scanning Electronic Microscopy with Electron Diffraction Spectroscopy (SEM-EDS) and, liquid samples obtained (condensate and liquified biomass) were analysed through Fourier Transformed InfraRed Spectroscopy (FTIR). The SEM microscope used was a model JSM-7001F (JEOL, Tokyo, Japan), where the solid samples were previously conducted through an gold alloy and, the FTIR spectrometer used was one from Agilent Technologies (Santa Clara, CA, USA), where the correspondent spectra were acquired in a range of wavenumber between 650 cm^{-1} and 4000 cm^{-1}, with a resolution of 32 scans.min^{-1}.

2.4. Preparation of the Different Electrolytes (NaOH Concentration and Biomass Content)

Several electrolytes were prepared with two different concentrations of sodium hydroxide (0.4 and 1.2 M) in demineralized aqueous solution, without any liquified biomass content. After that, four electrolytes were prepared, all of them with 1.2 M of sodium hydroxide concentration, but with different liquified biomass weight contents (5, 10, 12 and 15%), for a total volume of water and biomass in each experiment, equal to 100 mL. For all of these experiments, the different electrolytes were placed in the electrolyte admission tank and it was open the valve with connects the tank to the electrolyser, to filling it. Then, it was plugged the terminals of the electric feed supply to the electrolyser. All of these experiments were conducted for 2 hours, measuring in each 15 minutes, the electrolyte temperature (T), voltage applied (V), current intensity (I), gas volumetric flow and, its volumetric composition. In the beginning and in the end of each experiment, it was measured the pH and conductivity of the correspondent electrolyte.

2.5. Preparation of Acidified Heterogeneous Catalyst

Since the Y zeolite catalyst was supplied in basic form (NaY), it was necessary to convert it in acidic form (HY), using ionic exchange with a 2 M aqueous solution of ammonium nitrate, for 6 h in an oil heating bath, with stirring, at 80 °C, to convert first, into the ammonium form (NH_4Y). After this time, the final solution was filtered under vacuum filtration and the collected solid was dried in an oven, overnight (14 hours approximately), at 90 °C. In the next day, the dried solid was placed in a furnace oven, at 500 °C for 8 hours, with a gradient heat of 5 °C·min^{-1} releasing ammonia gas and adsorbed water, converting the zeolite from the ammonium form (NH_4Y) to the acidic form (HY).

2.6. Experiments in the Syngas/Methane Reactor

First, in order to estimate the liquified biomass apparent density or bulk density, which is a property of powders, granules, and other "divided" solids, or any other masses of corpuscular or particulate matter. It is defined as the mass of many particles of the material divided by the total volume they occupy. The total volume includes particle volume, inter-particle void volume, and internal pore volume [27]. This bulk density was quantified in the liquified biomass sample (A2) collected after the liquefaction process, supplied by one of our research partners. It was placed it in the reactor, 100 mL, and then, sealed the reactor inlet and placed it in an oil heating bath with magnetic stirring, for 4 hours, at different temperatures. The outlet reactor was connected to a condenser in order to collect the release condensate at liquid state, which was collected in a cylindrical tank. After 4 h, the liquified biomass was then weighted, to compare with its initial mass, before the correspondent experience.

The characterization and quantification of the syngas composition was conducted in a second round of experiences, namely the methane gas produced and the oxygen content after the reaction with the liquified biomass, with and without HY zeolite solid catalyst. In all of these experiments, it was used 100 mL of liquified biomass in the reactor, and also, it was used in the electrolyser, 0.4 M of sodium hydroxide aqueous solution. In the experiments, different weight percentages of catalyst were used, together with the liquified biomass, in the methane reactor. 1 g, 2 g and 4 g of HY zeolite were weighted and then mixed with the 100 mL of liquified biomass in this reactor, which corresponds, respectively, to, approximately, 0.9%, 1.8% and 3.6% of mass catalyst concentration, at four different temperatures (150, 200, 250 and 300 °C). All these experiments were carried out at 4 hours, measuring several parameters each 30 minutes, such as, temperature in the electrolyser, (T), voltage applied (V), current intensity (I), gas produced volumetric flow and its volumetric composition, at the outlet reactor, quantifying also, the condensate volume produced and, in the end of each experiment, the volume of liquified biomass, to compare it with the initial one.

3. Results and Discussion

3.1. Characterisation of the Supplied Different Biomass Samples

The solid biomass used as raw-material in the liquefaction process, supplied by a pulp industry Portuguese Company was cork and eucalyptus bark, with a typical elemental composition show in Table 1. It's possible to see that, those values are in accordance with similar ones published elsewhere [28].

Table 1. Elemental composition, humidity and heating values of cork/eucalyptus bark solid biomass samples.

Component	Used (%(w/w)) [1]	From Literature (Ligneous Biomass, % (w/w)) [28]		
C	46.0–49.0	44.0–53.0	Cellulose	30.0–50.0
H	5.30–5.70	5.50–6.50	Hemicellulose	15.0–35.0
O	42.0–47.5 [2]	38.0–49.0	Lignin	20.0–35.0
N	1.00–2.00	0.00–2.00	Ashes	0.20–8.00
S	0.08–1.00	0.05–1.00		
Cl	0.05–0.25	–		
Total humidity	44.0–67.0	Variable		
HHV (MJ/kg)	17.5–19.5	15.0–19.0		
LHV (MJ/kg)	16.5–18.5	–		

(1)—Obtained before entering in the hydro liquefaction process used, after several measures performed; (2)—Estimated by difference from total weight composition.

After the liquefaction process of these biomass samples, they were analyzed, through the quantification of its elemental composition, water content, the low heating value (LHV) and the high heating value (HHV). This analysis was performed in one of our research partners, a cement kiln producer and, the correspondent results are shown in Table 2.

Table 2. Elemental composition, humidity and heating values of cork/eucalyptus liquified biomass samples.

Component	% (w/w)	Component	% (w/w)
C	60.0–70.0	S	<0.50
H	12.0–13.0	Total humidity	2.00–4.00
O	14.5–25.5 [1]	HHV (MJ/kg)	31.5–39.0
N	<2.50	LHV (MJ/kg)	29.0–36.0

(1)—estimated by difference from total weight composition.

Table 2 shows elemental composition of liquified cork/eucalyptus bark biomass samples, after the hydro liquefaction process performed, in our research partner, as well, the final humidity content and the heating values (HHV and LHV).

As reported in several references, biomass solid samples shows higher H/C and O/C ratios then fossil fuels, like coal [28], which enhances hydrogen composition in the syngas production, in thermochemical process, like gasification [28]. In this electrochemical process, the increase in the hydrogen composition of syngas produced will enhance the production of biofuels, like in the methanation processes. Besides that, less carbon contents will decrease carbon dioxide emissions (GHG) to the atmosphere [28].

3.2. Preparation of the Different Electrolytes (NaOH Concentration and Biomass Content)

Table 3 shows the experimental results achieved regarding the first round of experiences in the electrolyser, with and without different liquified biomass contents, for two sodium hydroxide electrolyte concentrations. The parameters measured for each experiment were, the electrolyte temperature in the electrolyser (T), the input voltage (V), the current intensity (I), the final volumetric flow of gas produced, the initial and final pH (pHi and pHf), as well, the initial and final conductivity (Ki and Kf). All the values pointed in the table are averages values calculated from 4–5 experiences performed for each case, at the end of 120 min.

Table 3. Experimental results achieved regarding the experiences performed in the electrolyser.

Electrolyte	T (°C)	V (V)	I (A)	F (mL.min^{-1})	pH$_i$	pH$_f$	ΔpH (%)	K$_i$ (mS.cm^{-1})	K$_f$ (mS.cm^{-1})	ΔK (%)
NaOH 0.4 M	76.9	28.5	3.64	98.80	14.2	14.1	0.70	76.70	68.13	11.2
NaOH 1.2 M	53.2	28.5	2.50	103.0	14.2	14.2	0.00	198.8	194.9	1.96
NaOH 1.2 M + 5% (w/w) biom.	53.1	28.5	1.24	50.28	14.2	11.1	21.8	182.5	56.67	70.0
NaOH 1.2 M + 10% (w/w) biom	45.8	28.5	0.96	48.79	14.2	10.9	23.2	136.4	48.73	64.3
NaOH 1.2 M + 12% (w/w) biom	43.2	28.5	0.86	39.15	14.2	12.7	10.6	124.6	44.70	64.1
NaOH 1.2 M + 15% (w/w) biom	46.4	28.5	1.09	50.76	14.3	13.0	9.09	116.3	38.57	66.8

This table also shows the calculated correspondent relative variations of pH (ΔpH) and conductivity (ΔK). Through these results, it is possible to conclude that, with the exception of the 15% (w/w) of biomass content experiment, the increase in the liquified biomass in the electrolyte decreases the final temperature in the electrolyser and, also, the current intensity, as well, the produced gas volumetric flow, since, both parameters are related. The justification is directly related with the electrolysis conversion and the gas flow observed, which are proportional to the current intensity and, also, to the heat produced in the electrolyser, through Joule's effect. For the same electrolyte quantity, less current intensity will decrease the heat released and the electrolyser temperature. It's possible to see, also, that, the presence of organic compounds, such as, the liquified biomass, mixed in the electrolyte, decreases significantly the conductivity values. It's well known that, organic compounds have very low electrical conductivity values, thus affecting negatively the electrolysis conversion process, decreasing the electrolyte final conductivity, when compared with the sodium hydroxide conductivity values. This is confirmed by these results, which affected also, the final pH of the electrolyte, decreasing it.

In order to understand the evolution of the oxygen concentration in the electrolyser, over time, for different tested electrolytes (with and without liquified biomass), Figure 4 shows those evolutions. It is possible to notice that, higher biomass content mixed with the 1.2 M NaOH electrolyte will increase the oxygen consumption, due to its reaction with the carbon from biomass, thus producing CO and CO_2. It is the co-electrolysis processes, like reported elsewhere, by Guerra et al. [24,25]. It is, also, interesting to see that, A4 liquified biomass samples shows the same behavior as the 1.2 M NaOH aqueous electrolyte without any biomass content. This is due to the presence in this sample A4, of a large amount of solvent, mainly, water, ascribable to the high solubilization of the sugars compounds. On the contrary, the remaining biomass samples tested (A2) shows a strong oxygen conversion, because this sample only has a little portion of solvent, since it was previously removed in the liquefaction process.

For A2 experiments, it was also observed, a maximum oxygen conversion between 30 and 45 min, although the final oxygen content in the gas produced from the electrolyser, at 120 min, is higher, when compared with the observed at 30–45 min. This difference is due to the non-steady state process, which occurs until it finishes, 10–15 min after the 2 h of each experience performed.

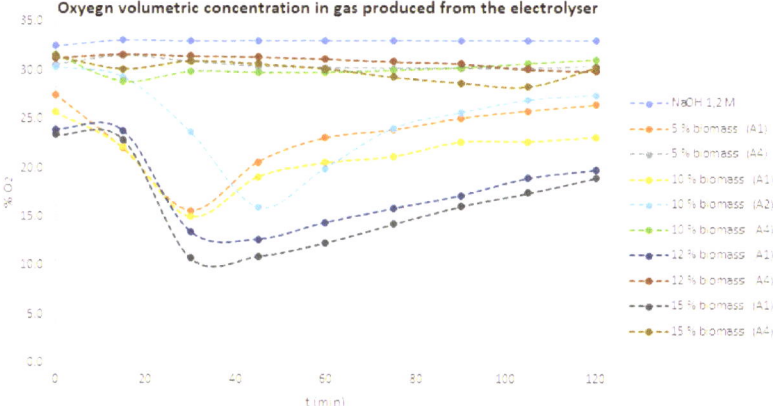

Figure 4. Evolution of the oxygen concentration in the electrolyser, over time, for different tested electrolytes (average values).

3.3. Experiments in the Syngas/Methane Reactor

Regarding the production of syngas, Table 4 shows the results obtained with complementary experiences in the methane glass reactor, where this gaseous mixture is produced, with the methane generation, through the reaction between the electrolysis gas and, the liquified biomass. The results achieved and calculated were: the volumetric yields production of solid, liquid and gas phases, respectively, the remaining biomass collected in the reactor (Yield$_{(liq.biom.)}$) after the time experience considered, at different temperatures, the condensate (Yield$_{(cond.)}$) and, also, the gas mixture produced (Yield$_{(gas)}$).

Table 4. Experimental results obtained with complementary experiences in the syngas/methane reactor.

Liquified Biomass	T (°C)	t (min.)	Vf$_{(liq.biom.)}$ (mL)	Yield$_{(liq.biom.)}$ (%)	V$_{(cond.)}$ (mL)	m$_{(cond.)}$ (g)	ρap. (cond.) (g·cm^{-3})	Yield$_{(cond.)}$ (%)	Yield$_{(gas)}$ (%) (*)
A2	100	60	98	98.0	-	-	-	-	2.0
	150	60	96	97.0	-	-	-	-	3.0
	200	30	-	-	18.5	18.2	0.98	18.5	-
		60	72	74.0	23.5	22.8	0.97	23.5	2.5
		240	74	72.0	24.5	23.5	0.96	24.5	3.5

(*)—estimated considering the initial volume of 100 mL of liquified biomass minus the volumes of final liquified biomass and condensate produced.

The apparent density of the liquid condensate (ρ$_{ap. (cond.)}$) was also calculated. The apparent density of the liquified biomass sample (A2) was previously calculated, giving an average value of 1.14 g·cm^{-3}. From these data, it is possible to conclude that, as expected, the increase of temperature will increase, at shorter times, the volume of produced condensate, thus decreasing the volumetric yield in the remaining liquified biomass. The calculated apparent density of the condensate is decreasing with the increase of the reactor temperature and, those values are similar with the water density, although, as explain more ahead, this condensate has, also, organic compounds, at minor concentrations.

Different operational parameters were studied in the syngas/methane reactor, such as the volumetric flow rate of gas produced over time (Q$_{vRnormalized}$), the volumetric percentage of oxygen produced (%O$_2$) and reacted over time (Q$_{vRO2consum}$) and, the volumetric percentage of methane produced (%CH$_4$). These tests were carried out with different weight contents of zeolite HY catalyst and different temperatures, like described in Section 2.6. The obtained results are presented in Figures 5–8. All these experimental rounds took place with the following fixed experimental conditions in the electrolyser: 22.4 V of applied voltage, 2.5 A of current intensity, 160 mL/min of electrolyser gas produced (H$_2$ + O$_2$), electrolyte of NaOH 0.4 M aqueous solution and, with 4 h in each experience.

Table 5 shows, at the end of 4 hours experience, the correspondent final output values of the gas volumetric flow, as well, the oxygen and methane volumetric contents in the produced gas mixture, for different reaction temperatures and different weight content (z. HY catalyst). To compare with another Y zeolite already prepared, ultra-stabilized with nickel (z. USY), it were also performed, two more experiments with this catalyst, which was supplied from another Portuguese university. The results achieved with USY zeolite doesn't show any significant improvement, mainly in the %CH_4 content, when compared with the acidified HY zeolite catalyst.

Table 5. Experimental results in the methanation reactor, for different reaction temperatures and different weight content catalyst.

Liquified Biomass	% ($W_{cat.}/W_{liq.biom.}$)	T (°C)	F (mL·min^{-1})	%O_2	%CH_4
		150	142.9	33.5	0.19
	—	200	150.0	32.0	0.45
		250	138.5	33.3	0.45
		300	138.5	32.0	2.08
		150	145.2	33.8	0.25
A2	z. HY, 2%	200	134.2	32.5	1.84
		250	145.2	32.2	4.16
		300	157.9	30.2	12.8
		150	145.2	33.9	0.28
	z. HY, 4%	200	145.2	32.5	3.98
		250	134.2	30.0	5.02
		300	138.5	22.2	33.9
	z. USY, 1%	200	125.0	33.1	0.17
	z. USY, 2%	200	132.4	33.1	0.26
A3	—	200	145.2	32.3	0.16
A4	z. HY, 4%	200	133.6	32.2	3.81

3.3.1. Flow of Produced Gases

Analyzing Figure 5, the observed flows exhibit the same general behavior. An exception is the flow rate for the test with 2% of HY catalyst at 300 °C, which was constant, due to a leak in the system. This leak did not affect the test at all, but only the flow measurement.

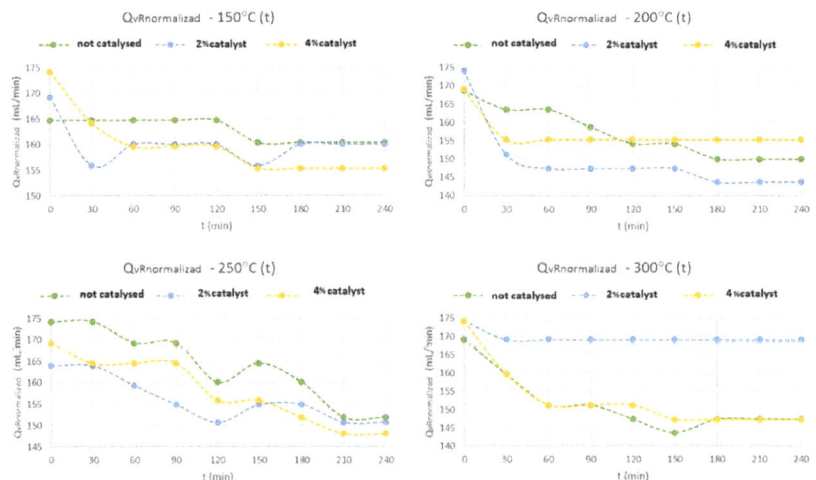

Figure 5. Comparison of the behavior of the volumetric flow rate of gases produced over the test time for the synthesis gas production tests for different amounts of catalyst and temperature (average values).

Nevertheless, the final gas flow measurements are basically constant, in all experiments, after 120–150 min of reaction time, when it was reached the steady state, with little differences (<10 mL·min^{-1}) between them, when it was reached the 4 hours of reaction time.

3.3.2. Oxygen Concentration in the Outlet Gas Mixture

In Figure 6, it can be seen that, the increase in temperature and the increase in the catalyst weight content affect the percentage of oxygen, i.e., for the temperatures of 250 and 300 °C and with 4% ($W_{catalyst}/W_{liq.biom.}$) of catalyst, it is possible to notice an appreciable decreasing on the oxygen concentration, in the syngas mixture, is more noticeable, after 210 min of reaction time.

At lower temperatures of 150 and 200 °C, the oxygen content only shows slight variations between 32.5 and 33.8% (v/v). It seems that, for temperatures above 250 °C and, with 4% of zeolite HY heterogeneous catalyst, the production of methane gas is enhanced, in the syngas mixture, because the activation energy boundary is being decreased. In all these experiments, slightly concentrations of CO and CO_2 were detected, which means that, the gas compound with more and significant concentration is the hydrogen.

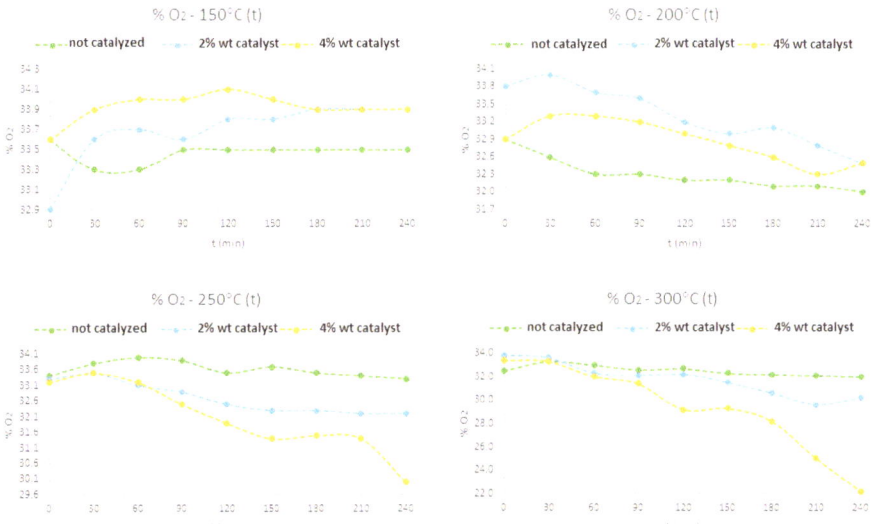

Figure 6. Comparison of the behavior of the percentage of oxygen in the gas produced over the test time for the synthesis gas production tests for different amounts of catalyst and temperatures (average values).

3.3.3. Oxygen Flow Rate Consumed

The oxygen consumption evolution is shown in Figure 7, where it is possible to observe the increase in the reacted oxygen volumetric flow with the increase of temperature, mainly at 250 and 300 °C and, with 4% of z. HY catalyst. At 150 °C, this consumption does not exceed 6 mL·min^{-1}, and, at 200 °C, this consumption has already reached 12 mL·min^{-1}. At 250 °C, this consumption increased to 14 mL·min^{-1}, and, at 300 °C, this consumption is even higher, reaching 28 mL·min^{-1}, for the test with 4% of z. HY catalyst. The decreasing on the oxygen content in the syngas mixture, will increase the reacted oxygen to produce, mainly, methane gas. These conditions are enhanced with the increase of reaction temperature above 250 °C and, also, with 4% of weighted z. HY catalyst, for the same reasons pointed out above.

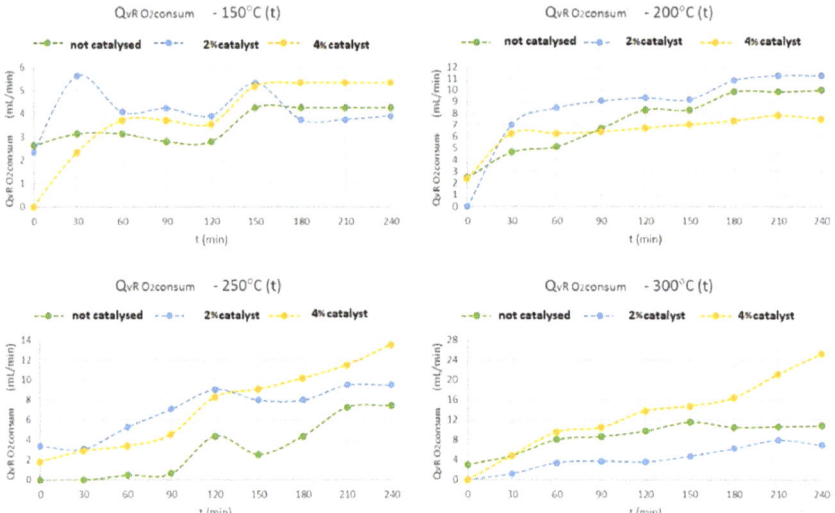

Figure 7. Comparison of the behavior of the volumetric flow rate of oxygen consumed over the test time for the synthesis gas production tests for different amounts of catalyst and temperatures (average values).

3.3.4. Methane Concentration

Figure 8 shows the production of methane through time, for different temperatures and different contents of zeolite HY catalyst, in the syngas/methanation reactor. At 150 °C, the methane concentration is very low, even with the use of the solid catalyst. Increasing the temperature and the amount of catalyst shows that, the methane production increases, with a maximum of 35%, obtained in the 300 °C test with 4% (w/w) of that catalyst. The reasons to explain this behavior of the methane production were explained above in this article, since the production of this fuel gas is directly related with the consumption of oxygen in this reactor, enhanced by the increase of temperature and, with, at least, 4% (w/w) of catalyst.

Best conditions which maximized methane concentration (300 °C, 4% (w/w)) were replicated three times and, the same behavior were observed, since, at the end of the 4 hour reaction time, final methane concentration achieved (yield) was 33% and 34% (twice), which gives an overage value of 34%, although, it was achieved also, a maximum concentration of 35% in all replicate experiments. The remaining gas compositions measured were basically the same. After these three replications, z. HY catalyst was calcinated again, to eliminate coal deposition in surface catalyst, to reactivate it, because coal deposition covered the catalyst active sites.

Another experiment with the same best operating conditions was performed, after catalyst recalcination in the same operating conditions and, the achieved results were the same of the previous ones, again with 34% of methane final concentration, at the end of 4 h of reaction time. After recalcination, catalyst acquire the same aspect as used in the first experiments.

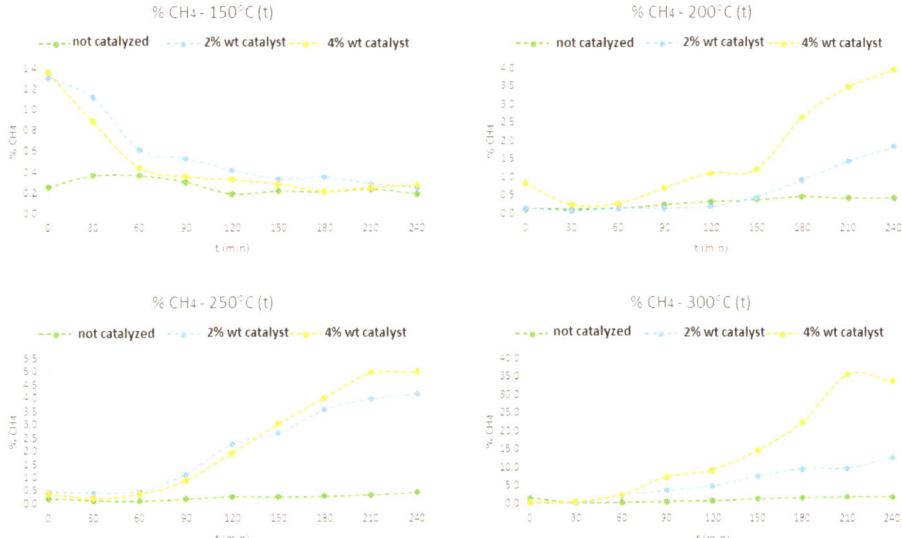

Figure 8. Comparison of the behavior of the percentage of methane in the gas produced over the test time for the synthesis gas production tests for different amounts of catalyst and temperature (average values).

In comparison with other similar studies, for instance, with Guerra et al. [24], it was reached a methane volumetric content of 25.7% of the syngas mixture produced in the 1 kW SYM electrochemical reactor, combined with a follow fixed bed catalytic reactor, under atmospheric pressure and, at 125 °C, using graphite electrodes in the electrolyser and, a Ni/(CaO-Al$_2$O$_3$) heterogeneous catalyst. These conditions give a methane gas selectivity of 96.5%, a CO$_2$ conversion of 44.2% and, residual concentrations of CO. In another study performed also, by the same authors, Guerra et al. [25] they achieved volumetric CO$_2$ concentration values of between 2.00–2.50% at 2 bar and 70 °C, but with residual values of methane gas and, with 25% of CO. In this last case, it isn't occurred any significant production of methane gas, which means, it doesn't show any relevant processual advantages when compared with the current study. No more similar studies were found in the literature, regarding the electrolytical production of syngas. Besides these two references, only pyrolysis/gasification process shows significant volumetric syngas and methane gas concentrations but, achieved with significant higher temperatures, higher than 400 °C and, in the case of pure syngas, only higher than 700 °C, in gasification process. These higher temperatures will need higher energy inputs for the syngas production, when compared with the electrochemical processes.

3.4. Stoichiometric Analysis

Due to the existence of methane in the final gas produced, it is apparent that the following main reactions occur [24,25], between carbon from biomass and the oxygen and hydrogen produced in the alkaline electrolyser. In the electrolyser itself, it occurs, both at the same time, the following reactions:

Anode:
Water alkaline oxidation:

$$4OH^- \xrightarrow{(45-55)\,°C} 2H_2O + 4e^- + O_2 \qquad (1)$$

Cathode:

Water alkaline reduction:

$$4H_2O + 4e^- \xrightarrow[(45-55)\,°C]{} 4OH^- + 2H_2 \qquad (2)$$

In the syngas/methane glass reactor, with the presence of the liquified biomass, attending the production of methane gas, the follow reactions seems to take place, according with [19,24,25]:

Partial carbon oxidation:

$$C + \tfrac{1}{2}O_2 \xrightarrow[(100-300)\,°C]{} CO \quad \Delta Hr = -222\;kJ\cdot mol^{-1} \qquad (3)$$

Total carbon oxidation:

$$C + O_2 \xrightarrow[(100-300)\,°C]{} CO_2 \quad \Delta Hr = -394\;kJ\cdot mol^{-1} \qquad (4)$$

After the production of CO and CO_2 gases, in the same methanation reactor, it will be produced methane gas, according with the following reactions, described elsewhere [19], which is enhanced by temperature and catalyst increases:

Sabatier reaction:

$$CO + 3H_2 \xleftrightarrow{cat.,(200-300)\,°C} CH_4 + H_2O \quad \Delta Hr = -206\;kJ\cdot mol^{-1} \qquad (5)$$

Water-gas shift reaction:

$$CO_2 + H_2 \xleftrightarrow{cat.,(200-300)\,°C} CO + H_2O \quad \Delta Hr = +41\;kJ\cdot mol^{-1} \qquad (6)$$

The overall reaction from these two (reactions (5) and (6)) leads to the following one:

$$CO_2 + 4H_2 \xleftrightarrow{cat.,(200-300)\,°C} CH_4 + 2H_2O \quad \Delta Hr = -165\;kJ\cdot mol^{-1} \qquad (7)$$

This means that, in the methanation reactor, the above reactions took placed in the follow order: first, the reactions (3) and (4), simultaneously, then the reactions (5) and (6), which, together, leads to the reaction (7), where, the standard specific enthalpy reaction is equal to $-206 + 41 = -165\;kJ.mol^{-1}$.

According to the stoichiometry of these reactions, it is possible to obtain some outputs such as: molar flow at the outlet of the electrolyser, molar flow of oxygen and hydrogen at the outlet of the electrolyser, molar flow at the exit of the syngas reactor, molar flow of oxygen, methane, hydrogen, carbon dioxide and carbon monoxide at the exit of these reactor, as well, the molar flow rate of oxygen, methane and hydrogen consumed. From these outputs, in the 300 °C test with 4% of z. HY catalyst, where there is a higher percentage of methane produced, and, since there is a portion of the flow produced that is not justified by the stoichiometry of these reactions, it means that, there are compounds formed in addition to those mentioned before, at the same time, in the reactor. These compounds may be hydrocarbons resulting from cracking processes of biomass itself, enhancing the methane production [29]. As reported in the literature, the propagation step mechanism of cracking paraffins leads, inevitably, to the co-production of methane gas [29].

On the other hand, the very low contents of CO and CO_2 observed, suggests that, through the temperature reaction and heterogeneous catalyst used conditions, these gases were basically consumed to produce methane, through the Sabatier process. Besides that, the most part of the liquified biomass in the reactor, was converted to liquid condensate.

In the same test with 300 °C and 4% (w/w) of z. HY catalyst, it was also observed, after the 240 min of reaction time, the deposition of small black particles, thus covering the catalyst surface, which could be ascribable to the deposition of coke particles, resulting most probably from the follow reaction (Equation (8)), which typically occurs on the methane conversion processes in the presence of steam water and/or oxygen, as well, in the gasification of coal and biomass [15]:

$$2CO \underset{}{\overset{cat.,(200-300)\ °C}{\longleftrightarrow}} C + CO_2 \quad \Delta Hr = -173\ kJ\cdot mol^{-1} \tag{8}$$

Although the global process is exothermic, due to the negative values of reaction enthalpies, it's necessary supply heat in order to achieve the desired temperature. The same procedure occurs in the thermochemical processes of syngas/methane production, like pyrolysis and gasification. Pyrolysis process starts at 400 °C and, syngas production in the gasification process normally occurs from temperatures higher than 700 °C. In order to calculate the theoretical supply heat to the correspondent process ($\overline{\Delta H}^T$), it's necessary to calculate the calorific values of syngas and methane produced in both cases (electrochemical/Sabatier combined process and, both thermochemical processes mentioned above), for the 300 °C achieved in this study and, for 400 °C and 700 °C, which normally occurs in the pyrolysis and in the gasification processes, respectively. To perform this task, it's necessary to apply the following expression (Equation (9)):

$$\overline{\Delta H}^T = \overline{Cp}\cdot \Delta T \tag{9}$$

where $\overline{Cp}$ is the mean specific calorific capacity between 25 °C and the temperature (T) used, ΔT is the difference of temperature between 25 °C (room temperature) and the operating one (T). The product $\overline{Cp}\cdot \Delta T$ corresponds to the specific calorific heat which is needed to supply for the process.

To calculate $\overline{Cp}$ values, is need, in first place, calculate each Cp for the correspondent component, between 25 °C and the operating temperature. With Equation (10), it's possible to calculate each (Cp_i) value, through the thermodynamic values of (a), (b), (c) and (d), which were collected in this study, from the literature (Himmelblau, [30]). These values are showed in Table 6. Equation (11) calculate the overall specific calorific capacity ($\overline{Cp}$) for the syngas/methane mixture at the reactor outlet, where (x_i) represents the correspondent volumetric composition of each gas component:

$$Cp_i = \frac{\int_{25\ °C}^{T}(a + b\cdot T + c\cdot T^2 + d\cdot T^3)dT}{(T - 25)} \tag{10}$$

$$\overline{Cp} = \sum_{i=1}^{n} x_i \cdot Cp_i \tag{11}$$

Table 6. Thermodynamic values (a,b,c,d) expressed in J·mol^{-1} °C^{-1}, of gas components, for application in equation 10 [30].

Compound	a	b	c	d
CO	28.95	4.11 × 10^{-3}	3.55 × 10^{-6}	−2.22 × 10^{-9}
CO$_2$	36.11	4.23 × 10^{-2}	−2.89 × 10^{-5}	7.46 × 10^{-9}
O$_2$	29.10	1.16 × 10^{-2}	−6.08 × 10^{-6}	1.31 × 10^{-9}
N$_2$	29.10	2.20 × 10^{-3}	5.72 × 10^{-3}	−2.87 × 10^{-9}
H$_2$	28.84	7.65 × 10^{-5}	3.29 × 10^{-6}	−8.70 × 10^{-10}
CH$_4$	34.31	5.47 × 10^{-2}	3.66 × 10^{-6}	−1.10 × 10^{-8}

The values of ($\overline{Cp}$) for each case, depending of the operating temperature (T), applied in Equation (9), gives the values of specific heats. Table 7 shows those values for the analyzed processes. N$_2$ was only applied for outlet gases in pyrolysis and gasification processes, while O$_2$ only in the electrolytic process.

Table 7. Calculated values of $(\overline{Cp})$ and $(\overline{\Delta H}^T)$ for different syngas/methane production processes.

Process	T (°C)	$(\overline{Cp})$ (J.mol^{-1}.°C^{-1})	$\overline{\Delta H}^T$ (kJ · mol^{-1})
Electrochemical/Sabatier combination	300	34.62	10.38
Pyrolysis (with N_2 and without O_2)	400	35.39	14.15
Gasification (with N_2 and without O_2)	700	38.46	26.92

It's possible to conclude, according with $\overline{\Delta H}^T$ values that, higher temperature process means a significant increase in the input energy and, as consequence, a significant increase with the energetic (operating) costs. Comparing the pyrolysis process (14.15 kJ·mol^{-1}) with the combined electrochemical/Sabatier one (10.38 kJ·mol^{-1}) and considering the same syngas/methane flow and the same gas composition, an decrease of 36% in the input energy was observed. By another hand, the comparison between the same combined process (10.38 kJ·mol^{-1}) with the gasification one (26.92 kJ·mol^{-1}), an decrease of 159% in the input energy was observed, both values applied for each mole of syngas/methane mixture.

3.5. FTIR Analysis

Infrared spectroscopy analyzes were performed on some liquid samples obtained in the previous experiments. These liquid samples refer to the biomass used in the tests, before and after those tests, as well, in the condensate obtained. Analyzing Figure 9A,B, it can be seen that, the liquified biomass spectra are identical, before and after the trials, respectively, since the absorption peaks detected were almost the same, varying only their intensity.

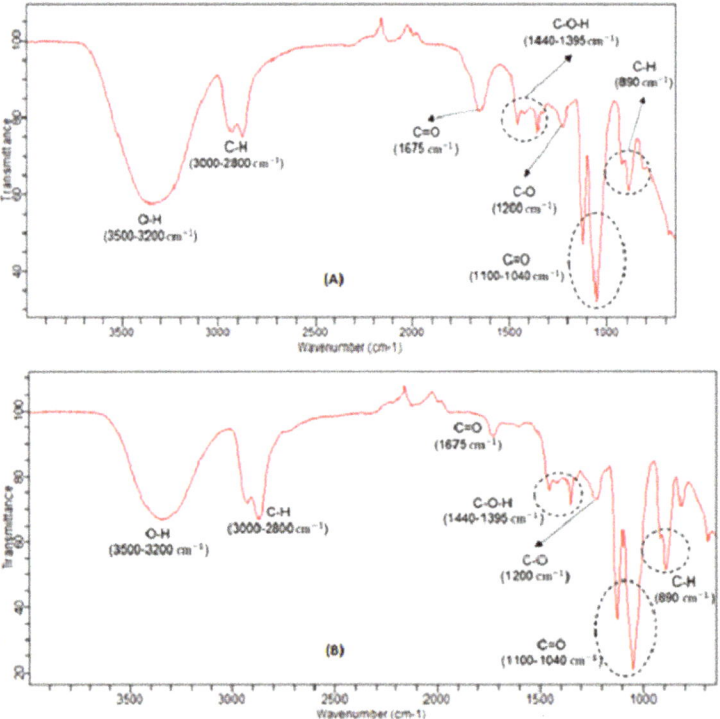

Figure 9. FTIR spectra: (A) corresponding to the initial sample of liquefied biomass, prior to any test; (B) corresponding to the biomass sample after the test at 250 °C, without catalyst.

It can be seen that, the most intensity peak is related with the O-H absorption peaks between 3200 and 3500 cm^{-1}, mainly ascribable with alcohols, water and, for the case of biomass samples, are related also with O-H bonds of the hydroxyl groups present in the several monomers of the cellulose and hemicellulose structures [31]. Nevertheless, it's possible to see a little decreasing in intensity of these absorption peaks of O-H (3200–3500 cm^{-1}), and also, at 1675 cm^{-1} related with C=O bonds (stretching vibrations) of aldehydes and ketones, after the correspondent experiment, due to the evaporation of some of these compounds to the condensate.

The spectra presented in Figure 10A,B are related with the liquid condensate samples obtained, respectively, for the trials of 150 °C with 4% of catalyst and, at 300 °C with 2% of catalyst. The remaining trials performed give similar condensate FTIR spectra to these two cases. For low reaction temperatures, it's possible to detect the O-H stretching vibrations bonds, typical in water, alcohols and similar compounds, but also, the C=O absorption peaks, at 1675 cm^{-1}, typical of aldehydes and ketones. For higher temperatures (250 and 300 °C), it is also possible to detect, besides these absorption peaks mentioned before, other ones, mainly at, 2800–3000 cm^{-1} and at 890 cm^{-1}, which correspond to the stretching and bending vibrations of the C-H bonds of the aldehydes, as well, at 1040–1100 cm^{-1}, for stretching vibrations of the C=O bonds of the carboxylic acids, 1200 cm^{-1}, addressed to stretching vibrations of the C-O bonds of the alcohols and, finally, at 1395–1440 cm^{-1}, ascribable to stretching vibrations of the C-O-H bonds of the carboxylic acids, also.

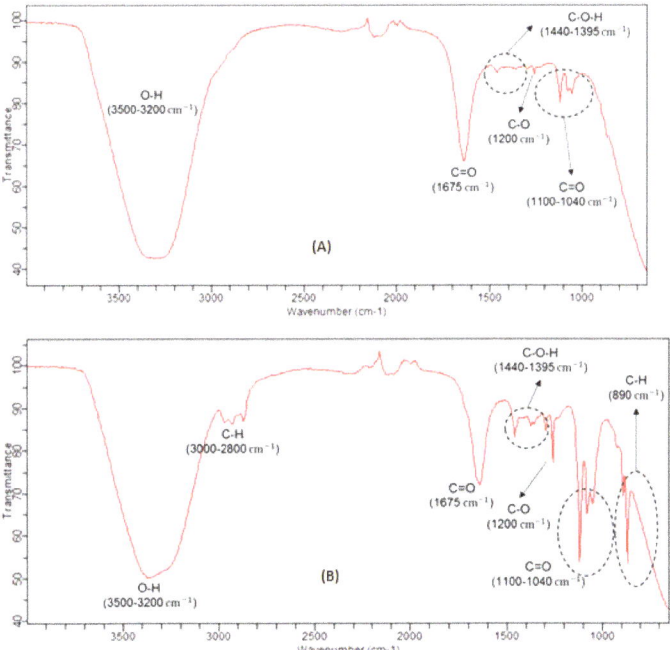

Figure 10. FTIR spectra: (**A**) corresponding to the condensate resulting from the test at 150 °C with 4%g of z. HY catalyst; (**B**) corresponding to the condensate resulting from the test at 300 °C with 2% of z. HY catalyst.

When comparing the FTIR spectra of liquified biomass and condensate liquid samples, it's possible to verify that, the functional groups which are decreasing its intensity in the biomass samples, increase in the condensate samples.

This fact was to be expected, since the most volatile constituents with O-H, C-O and C=O bounds evaporate during the reaction and, therefore, are collected in condensate tank. Since the evaporation of the condensate previous collected was almost complete at 100 °C, suggests that the major quantity of those alcohols, aldehydes and carboxylic acids have boiling temperatures below 100 °C, which could be ascribable to formaldehyde, methanol, ethanol and, formic acid, since these compounds have, all, normal boiling temperatures below than 100 °C.

3.6. SEM-EDS Analysis

By the end of the syngas production test at 300 °C, with 4 g of z. HY catalyst, a solid was obtained, with some black particles, ascribable to coke deposition. In order to observe and characterize morphologically, this sample, before and after the acidification and calcination processes, as well, after the reaction at those conditions, SEM-EDS analysis was used, as shown in Figures 11 and 12 (SEM images) and Figures 13 and 14 (EDS spectra – atomic percentages).

Figure 11. SEM images of z. HY catalyst sample collected after the methanation reaction process, at 300 °C and with 4% (w/w) of z. HY catalyst (**A**: 2000 x, 10 µm; **B**: 5000 x, 1 µm).

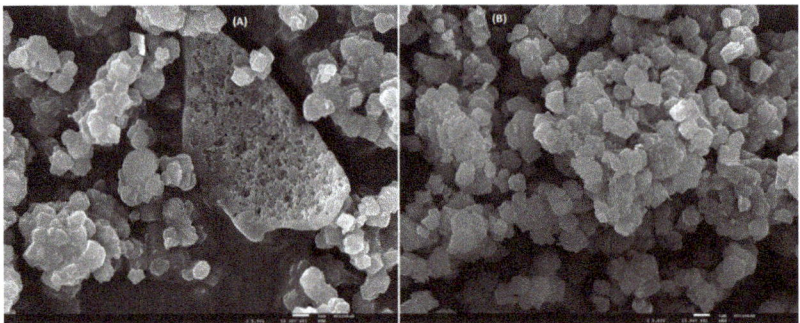

Figure 12. SEM images (5000 x; 1 µm) of: (**A**) z. HY catalyst sample collected before the activation process; (**B**) same catalyst sample collected after the activation process.

Analyzing the several SEM micrographics of these figures, it is possible to conclude that the solid sample is not homogeneous in its constitution, mainly in the solid sample collected after the reaction process, at 300 °C with 4% of weighted catalyst. The grey areas of post-reaction catalyst, according with SEM image of Figure 11B can be ascribable to the carbonaceous residue deposited on the surface catalyst, with a significant content, since the carbon atomic content increase significantly, from 3.4% to 68.1% and 76.2%, in two different points of the solid surface analyzed, according with Table 8.

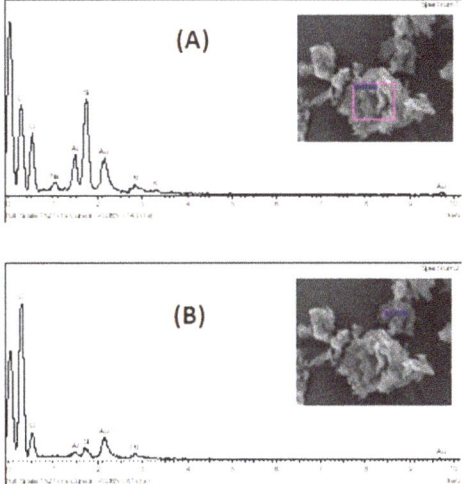

Figure 13. EDS spectra and respective SEM image of: (**A**) solid sample collected before the syngas production test at 300 °C with 4% of z. HY catalyst and (**B**) solid sample collected after the syngas production test at 300 °C with 4%g of z. HY catalyst.

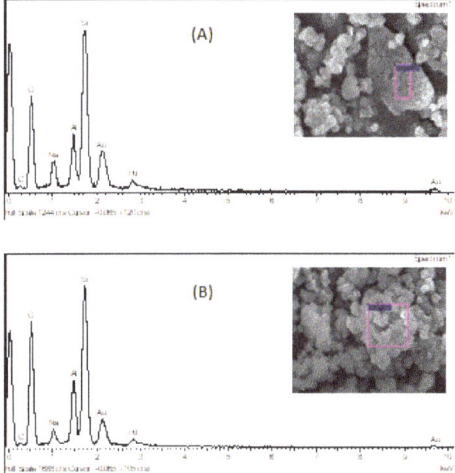

Figure 14. EDS spectra of: (**A**) z. HY catalyst sample collected before the activation process and (**B**) same catalyst sample collected after the activation process.

Several researchers pointed out in their articles that, this coke can be eliminated to CO_2, regenerating the zeolite HY catalyst, to be active again in this reaction. The same phenomena was also observed in this work. It is well known that, this catalyst is the same that is currently used in the fluid catalytic cracking of heavy diesel fuels, in the crude oil refining industry [32]. The catalyst has the same behavior, in both processes.

Besides this, when comparing SEM images and EDS analysis of atomic contents, before and after the catalyst activation process (acidification and calcination), there aren't significant changes in the morphology of the catalyst surface, with a little exception in the decreasing of the sodium content, which was expectable, due to the ionic exchange performed, where the sodium cation was leached.

The carbon atomic content decreased also, probably due to the calcination process, where the adsorbed CO_2 was released of the catalyst surface, to the atmosphere and/or, due to the decomposition of some sodium carbonate adsorbed, which was converted to sodium oxide with CO_2 released.

Table 8. Atomic percent data obtained by EDS spectra for catalyst samples, before and after the activation processes, as well, after the biomethane reaction, at 300 °C with 4% (w/w) of zeolite HY.

Element	Before Activation	After Activation	After Reaction	
			Sample 1 (Spectrum 1)	Sample 2 (Spectrum 2)
C	6.57	3.43	68.1	76.2
O	57.5	64.4	24.8	22.7
Na	5.29	2.29	0.63	-
Al	6.73	7.74	1.58	0.36
Si	24.0	22.1	4.74	0.77
K	-	-	0.18	-

4. Conclusions

From this research work, it can be concluded that it is possible to produce syngas and methane, using this electrolysis system (electrofuel), together with a fixed bed catalytic reactor to produce methanation (Sabatier process), with significant less energy inputs when compared with the conventional thermochemical processes of syngas/methane production, like pyrolysis and gasification. Comparing the combined electrochemical/Sabatier process (10.38 kJ·mol^{-1}) with the pyrolysis one (14.29 kJ·mol^{-1}) and considering the same syngas/methane flow and the same gas composition, an increase of 38% in the input energy was observed. By another hand, the comparison between the same combined process (10.38 kJ·mol^{-1}) with the gasification one (27.21 kJ·mol^{-1}), an increase of 162% in the input energy was observed, both values applied for each mole of syngas/methane mixture. With the utilization of this combined electrochemical/Sabatier reactors, it's possible to reduce input energy to the system and, as consequence, reduce energetic (operating) costs.

Regarding the methane production in this reactor, the operating conditions obtained so far, which enhanced and maximized its production was, a temperature of 300 °C and a weight heterogeneous catalyst content of 4% of zeolite HY. However, it should be noticed that, there are compounds, in the produced gas, that were measurable by the portable sensors. It was possible to conclude also, that, z. HY catalyst was progressively deactivated, through the visualization of carbon particles deposition on the surface catalyst. Nevertheless, the catalyst can be reactivated, by calcination, to be used again in the Sabatier reaction, so it's possible to conclude that, the use of z. HY catalyst was clearly suitable in the Sabatier reaction (methanation process), at normal pressure and temperatures between 200–300 °C.

Besides, the use of acidified zeolite HY catalyst and higher temperatures increases methane production, which points out for further research steps comprising the increase of catalyst mass, and, to study the increase of pressure and temperature in a new laboratory prototype. It will also be of interest to investigate the use of other heterogeneous catalysts which may be more active such as other zeolites, acid clays or bimetallic catalysts, as well, study the production of other biofuels, like biomethanol, bio-DME, etc., regarding this electrolytic system.

Author Contributions: Conceptualization, J.F.P. and J.F.G.; Execution and discussion of the results, A.G., L.G. and J.C.R.; writing—review and editing, M.T.S. and D.A.

Funding: This research was funded by F.C.T. (Fundação para a Ciência e Tecnologia), grant number SAICT-POL/23470/2016.

Acknowledgments: The authors thank SECIL, S.A. and Margarida Mateus (Ph.D. researcher) for supplying the liquefied cork samples used in this experimental work. The authors also thank [a] Maria Celeste Serra, from CEEQ/ISEL, for equipment utilisation, [a] Isabel Nogueira from MicroLab, at IST-UL, for SEM-EDS analysis, [a] Ana Ribeiro, from CQE/IST-UL, for FTIR utilisation and, Carlos Henriques, also from CQE/IST-UL, for supplying the USY zeolite catalyst.

Conflicts of Interest: The authors declare no conflict of interest.

Nomenclature

%CH$_4$	volumetric percentage of methane produced
%O$_2$	volumetric percentage of methane produced
ΔHr	standard specific reaction enthalpy, at 1 atm and 25 °C
$\overline{\Delta H}^T$	specific heat (specific calorific value) of syngas/methane mixture at T temperature
ΔT	difference of temperature = T − 25
A1, A2, A3, A4	Liquified biomass samples codification
$\overline{Cp}$	mean specific calorific capacity for syngas/methane mixture
Cp$_i$	specific calorific capacity for each gas component
FTIR	Fourier Transformed InfraRed Spectroscopy
GHG	greenhouse gas
HHV	high heating value
LHV	low heating value
Q$_{vRnormalized}$	volumetric flow rate of produced gas
Q$_{vRO2consum}$	volumetric flow rate of oxygen consumed
SEM-EDS	Scanning Electronic Microscopy with Electron Diffraction Spectroscopy
W$_{catalyst}$/W$_{liq.biom.}$	Weight catalyst content regarding weight of liquified biomass employed
T	temperature
z. HY	acidified zeolite HY catalyst
z. USY	ultra-stabilized zeolite Y catalyst

References

1. IEA. IEA 2017. Available online: www.iea.org (accessed on 26 August 2019).
2. BP. Statistical Review of World Energy, June 2018, BP. Available online: www.bp.com (accessed on 26 August 2019).
3. Atabani, A.; Silitonga, A.; Badruddin, I.; Mahlia, T.; Masjuki, H.; Mekhilef, S. A comprehensive review on biodiesel as an alternative energy resource and its characteristics. *Renew. Sustain. Energy Rev.* **2012**, *16*, 2070–2093. [CrossRef]
4. Mohr, S.; Wang, J.; Ellem, G.; Ward, J.; Giurco, D. Projection of world fossil fuels by country. *Fuel* **2015**, *141*, 120–135. [CrossRef]
5. Lazkano, I.; NøStbakken, L.; Pelli, M. From Fossil Fuels to Renewables: The Role of Electricity Storage. *Eur. Econ. Rev.* **2017**, *99*, 113–129. [CrossRef]
6. Mateus, M.; Bordado, J.; Santos, R. Potential biofuel from liquefied cork—Higher heating value comparison. *Fuel* **2016**, *174*, 114–117. [CrossRef]
7. Huang, H.; Yuan, X. Recent progress in the direct liquefaction of typical biomass. *Prog. Energy Combust. Sci.* **2015**, *49*, 59–80. [CrossRef]
8. Gollakota, A.; Kishore, N.; Gu, S. A review on hydrothermal liquefaction of biomass. *Renew. Sustain. Energy Rev.* **2018**, *81*, 1378–1392. [CrossRef]
9. Haung, P.; Koj, M.; Turek, T. Influence of process conditions on gas purity in alkaline water electrolysis. *Int. J. Hydrog. Energy* **2017**, *42*, 9406–9418.
10. Durak, H. Bio-oil production via subcritical hydrothermal liquefaction of biomass. In International Conference on Advances in Natural and Applied Sciences. *Am. Inst. Phys.* **2017**, *1833*, 1–5.
11. Sapountzi, F.; Gracia, J.; Weststrate, C.; Fredriksson, H.; Niemantsverdriet, J. Electrocatalysts for the generation of hydrogen, oxygen and synthesis gas. *Prog. Energy Combust. Sci.* **2017**, *58*, 1–35. [CrossRef]
12. Carmo, M.; Fritz, D.; Mergel, J.; Stolten, D. A comprehensive review on PEM water electrolysis. *Int. J. Hydrog. Energy* **2013**, *38*, 4901–4934. [CrossRef]
13. Vincent, I.; Bessarabov, D. Low cost Hydrogen production by anion Exchange membrane electrolysis: A review. *Renew. Sustain. Energy Rev.* **2018**, *81*, 1690–1704. [CrossRef]
14. Nguyen, V.; Blum, L. Syngas and Synfuels from H$_2$O and CO$_2$: Current Status. *Chem. Ingénieur Tech.* **2015**, *87*, 354–375. [CrossRef]

15. Moulijn, J.; Makkee, M.; Diepen, A. *Chemical Process Technology*, 2nd ed.; John Wiley & Sons Ltd.: Bognor Regis, UK, 2013; ISBN 978-1118570746.
16. Bellotti, D.; Rivarolo, M.; Magistri, L.; Massardo, A. Feasibility study of methanol production plant from hydrogen and captured carbon dioxide. *J. CO2 Util.* **2017**, *21*, 132–138. [CrossRef]
17. Azizi, Z.; Rezaeimanesh, M.; Tohidian, T.; Rahimpour, M. Dimethyl ether: A review of technologies and production challenges. *Chem. Eng. Process. Process. Intensif.* **2014**, *82*, 150–172. [CrossRef]
18. Khodakov, A.; Chu, W.; Fongarland, P. Advances in the Development of Novel Cobalt Fischer-Tropsch Catalysts for Synthesis of Long-Chain Hydrocarbons and Clean Fuels. *Chem. Rev.* **2017**, *107*, 1692–1744. [CrossRef] [PubMed]
19. Stangeland, K.; Kalai, D.; Li, H.; Yu, Z. CO_2 methanation: The effect of catalysts and reaction conditions. *Energy Proc.* **2017**, *105*, 2022–2027. [CrossRef]
20. Loosdrecht, J.; Botes, F.; Ciobîcă Ferreira, A.; Gibson, P.; Moodley, D.; Saib, A.; Visagie, J.; Weststrate, C.; Niemantsverdriet, H. Fischer-Tropsh Synthesis: Catalysts and Chemistry. In *Comprehensive Inorganic Chemistry II*, 2nd ed.; Elsevier: Amsterdam, The Netherlands, 2013; Volume 7, pp. 525–557. ISBN 978-0-08-096529-1.
21. Krylova, A. Products of the Fischer-Tropsch Synthesis A Review. *Solid Fuel Chem.* **2014**, *48*, 22–35. [CrossRef]
22. Guerra, L.; Gomes, J.F.; Puna, J.F.; Rodrigues, J. Preliminary study of synthesis gas production from water electrolysis, using the ELECTROFUEL®concept. *Energy* **2015**, *89*, 1050–1056. [CrossRef]
23. Chen, X.; Guan, C.; Xiao, G.; Du, X.; Wang, J. Syngas production by high temperature steam/CO_2 co-electrolysis using solid oxide electrolysis cells. *Faraday Discuss.* **2015**, *182*, 341–351. [CrossRef] [PubMed]
24. Guerra, L.; Rossi, S.; Rodrigues, J.; Gomes, J.F.; Puna, J.F.; Santos, M.T. Methane production by a combined Sabatier reaction/water electrolysis process. *J. Environ. Chem. Eng.* **2018**, *6*, 671–676. [CrossRef]
25. Guerra, L.; Moura, K.; Rodrigues, J.; Gomes, J.; Puna, J.; Santos, M.T. Synthesis gas production from water electrolysis, using the Electrocracking concept. *J. Environ. Chem. Eng.* **2018**, *6*, 604–609. [CrossRef]
26. Gonçalves, A. Utilização de biomassa liquefeita na produção eletrolítica de gás de síntese. In *Master Thesis in Chemical and Biological Engineering*; ISEL: Lisboa, Portugal, 2017. (In Portuguese)
27. Buckman, H.O.; Brady, N.C. *The Nature and Property of Soils—A College Text of Edaphology*, 6th ed.; Macmillan Publishers: New York, NY, USA, 1960.
28. Dimitriadis, A.; Bezergianni, S. Hydrothermal liquefaction of various biomass and waste feedstocks for biocrude production: A state-of-the-art review. *Renew Sustain. Energy Rev.* **2017**, *68*, 113–125. [CrossRef]
29. Gary, J.; Handwerk, H.; Glenn, E. *Petroleum Refining—Technology and Economics*; Marcel Dekker Inc.: New York, NY, USA, 2001.
30. Himmelblau, D.M. *Basic Principles and Calculations in Chemical Engineering*, 6th ed.; Prentice Hall: New York, NY, USA, 1996.
31. Dhyani, V.; Bhaskar, T. A comprehensive review on the pyrolysis of lignocellulosic biomass. *Renew. Energy* **2018**, *129*, 695–716. [CrossRef]
32. Alam, S.; Bangash, S.; Bangash, F. Elemental Analysis of Activated Carbon by EDS Spectrophotometry and X-rays Diffraction. *J. Chem. Soc. Pak.* **2009**, *31*, 46–58.

© 2019 by the authors. Licensee MDPI, Basel, Switzerland. This article is an open access article distributed under the terms and conditions of the Creative Commons Attribution (CC BY) license (http://creativecommons.org/licenses/by/4.0/).

Article

Techno-Economic Performance of Different Technological Based Bio-Refineries for Biofuel Production

Shemelis N. Gebremariam [1], Trine Hvoslef-Eide [2], Meseret T. Terfa [3] and Jorge M. Marchetti [1,*]

1. Faculty of Science and Technology (REALTEK), Norwegian University of Life Sciences, Drøbakveien 31, 1432 Ås, Norway; Shemelis.nigatu.gebremariam@nmbu.no
2. Department of Plant Sciences (IPV), Faculty of Biosciences, Norwegian University of Life Sciences, Arboretveien 6, 1432 Ås, Norway; trine.hvoslef-eide@nmbu.no
3. Department of Plant & Horticultural Sciences, College of Agriculture, Hawassa University, P. O. Box 05, Hawassa, Ethiopia; mesitesema@gmail.com
* Correspondence: jorge.mario.marchetti@nmbu.no

Received: 28 July 2019; Accepted: 11 October 2019; Published: 16 October 2019

Abstract: There are different technologies for biodiesel production, each having its benefits and drawbacks depending on the type of feedstock and catalyst used. In this study, the techno-economic performances of four catalyst technologies were investigated. The catalysts were bulk calcium oxide (CaO), enzyme, nano-calcium oxide, and ionic liquid. The study was mainly based on process simulations designed using Aspen Plus and SuperPro software. The quantity and quality of biodiesel and glycerol, as well as the amount of biodiesel per amount of feedstock, were the parameters to evaluate technical performances. The parameters for economic performances were total investment cost, unit production cost, net present value (NPV), internal return rate (IRR), and return over investment (ROI). Technically, all the studied options provided fuel quality biodiesel and high purity glycerol. However, under the assumed market scenario, the process using bulk CaO catalyst was more economically feasible and tolerable to the change in market values of major inputs and outputs. On the contrary, the enzyme catalyst option was very expensive and economically infeasible for all considered ranges of cost of feedstock and product. The result of this study could be used as a basis to do detail estimates for the practical implementation of the efficient process.

Keywords: biodiesel; CaO catalyst; nano-catalyst; ionic liquid catalyst; economic analysis

1. Introduction

According to the recent report from the World Energy Outlook 2018 [1], 93% of the world's carbon capacity is already in use up to 2040. Consequently, there is a very narrow space for the development of fossil fuel projects over this period without contradicting international objectives about climate change. This implies that it is becoming inevitable to push on the development of alternative and renewable energy resources for the supply of reliable and environmentally efficient energy to the growing economic activities around the world. Among such alternative sources are biofuels [2], which are mainly preferred for their carbon neutral character, their renewability, as well as the fact that they can be produced in decentralized manners from abundant and versatile resources. Biodiesel is one of the promising biofuels to substitute conventional fossil diesel. It has a number of environmental and technical benefits over conventional fossil diesel. Environmentally, biodiesel is non-toxic, biodegradable, and its greenhouse gas (GHG) emission is very low compared to the conventional fossil diesel [3,4]. The technical benefits are associated with its use for fuel; for example, its possession of more oxygen to favor complete combustion and its better lubricating character to reduce engine wear [5].

Biodiesel can be produced from different oil and fat resources, which are found everywhere. Such feedstocks include edible and non-edible plant oil, animal fat, as well as waste oils and fats. The production of fuel quality biodiesel from oil and fat feedstock mainly involves the transesterification reaction with alcohol in the presence of some kind of catalyst or without catalyst at the supercritical condition. The transesterification reaction catalyzed by homogeneous base catalysts like NaOH and KOH is the conventional way of producing biodiesel at an industrial scale [6], which requires relatively better-quality feedstock like edible oil with very low free fatty acid (FFA) content [7,8]. Such high-quality feedstock is usually associated with a high price. In addition, it creates food versus energy controversies. These reasons altogether have been making biodiesel the expensive alternative fuel compared to its counterpart-fossil diesel, because the cost of feedstock can take up to 80% of the total cost of biodiesel production [9,10]. Comparatively, the heterogeneous alkali-catalyzed transesterification reaction has advantages of easy catalyst recovery and reusability for multiple times [11,12]. Unlike the homogeneous ones, the heterogeneous alkali-catalyzed transesterification can tolerate a considerable amount of FFA in the feedstock. For instance, Avhad et al. [13] reported that using glycerol-enriched calcium oxide as a heterogeneous alkali catalyst, a 96.1% of crude *Jatropha curcas* oil containing high free fatty acid could be converted into biodiesel within 7 h. In addition, most of such heterogeneous base catalyst types can be easily prepared from cheap resources, indicating the potential to reduce biodiesel production cost. For example, industrial wastes (red mud, slag, ash) and biological wastes (chicken eggshells, mollusk shells, animal bones) have huge potential towards developing a cheap catalyst for low-cost biodiesel production [14]. Among the heterogeneous alkali catalysts developed for biodiesel production, the main ones include basic zeolites, alkaline earth metal oxides, and hydrotalcites [15].

The conventional chemical catalyst options also include homogeneous and heterogeneous acid catalysts. In general, acid-catalyzed transesterification is very efficient in the production of biodiesel from feedstock with very high FFA content [16,17]. However, the problems usually associated with the use of acid catalysts are high reaction temperature, longer reaction time, and corrosion of the equipment due to the acid catalyst [10]. There are some substantial advantages of solid acid catalysts over the homogeneous ones. This includes ease of catalyst separation from the reaction media, which lowers product contamination and ease of catalyst regeneration and reuse, as well as much-minimized equipment corrosion [18].

The other most promising technologies for the production of biodiesel from least cost feedstock involve CaO-based catalysts, enzyme catalysts, ionic liquid (IL) catalysts, and nanoparticle catalysts. Calcium oxide, as a catalyst, has such advantages as an abundance occurrence, better catalytic property, easy separation from the product stream, reusability for multiple times, nontoxicity, and least cost character for feasible production of biodiesel from lower-quality feedstock [19,20]. Boey et al. [21] did a study on the production of biodiesel from waste cooking oil using a CaO catalyst derived from waste sources like mud crab shells and cockleshells. They calcined the CaO obtained from these wastes at 900 °C for 2 h separately and mixed them in a 1:1 mass ratio to catalyze the transesterification of the oil. According to their result, a 98% conversion could be achieved within 3 h for optimum reaction conditions of 5 wt.% catalyst and methanol to oil molar ratio of 13:1 at methanol refluxing temperature [21]. In another study, Sasiprapha et al. [22] assessed the production of biodiesel from used oil using CaO catalyst derived from river snail shells. For optimum reaction conditions of methanol to oil ratio of 9:1, catalyst amount of 3 wt.%, and reaction temperature of 65 °C, they could achieve 92.5% conversion of oil to biodiesel within 3 h [22].

Even though the enzyme-catalyzed approach for biodiesel production is the expensive option, primarily due to the cost of the enzyme, the technical performance of most enzyme catalysts for the production of fuel quality biodiesel is very significant. Enzyme for the catalysis of biodiesel production has such advantages over the chemical catalysts as being less energy-intensive, allowing easy recovery of glycerol from the product stream and efficient conversion of acidic oil (oil with high FFA content) to biodiesel [23,24]. A study done by Cervero' et al. [25] indicated that a 95% conversion of soybean oil to biodiesel could be reached within 24 h using Novozyme 435 enzyme catalyst at optimum reaction conditions of 5 wt.% enzyme load, 3:1 molar ratio of ethanol to oil, and a temperature of 37 °C. Ketsara et al. [26] also studied the production of biodiesel from used palm oil using mixed enzymes in a solvent-free environment. The studied mixed enzyme contained *Pseudomonas fluorescens* and *Candida rugose*. According to their result, 89% conversion could be realized within 12 h for optimum reaction conditions of 3:1 ethanol to oil molar ratio, 10% enzyme dosage, 2% water content of the oil feedstock, and a 45 °C reaction temperature [26].

The other group of promising catalysts for biodiesel production are ionic liquids, which are generally known as solvents and green catalysts in chemical processes. Several ionic liquids are being used for the catalysis of biodiesel production from various low-cost feedstock alternatives. The use of such ionic liquids for biodiesel production provides considerable advantages over most other catalyst categories. Some of such advantages are low corrosion of equipment, ease of separation, recyclability, and less wastewater production [27]. In addition, the lower reaction time, together with the ability to produce good quality biodiesel from low-cost feedstock, could make ionic liquid catalysis a better alternative than most of the catalyst options for biodiesel production. Feng et al. [28] studied the transesterification process to produce biodiesel from palm oil using Brønsted acidic ionic liquid as a catalyst. They found out that conversion of 98.7% of the oil to biodiesel could be achieved within 2.5 h when the optimum reaction conditions are methanol to oil molar ratio of 21:1, catalyst dosage of 3 wt.%, and reaction temperature of 120 °C [28]. In another study, Ullah et al. [29] investigated the production of biodiesel from waste palm cooking oil using the acidic ionic liquid as a catalyst. They used specific ionic liquid butyl-methyl imidazolium hydrogen sulfate ($BMIMHSO_4$) as catalyst, and the highest biodiesel yield of 95.6% could be achieved with optimum reaction conditions of 5 wt.% of $BMIMHSO_4$, methanol to oil molar ratio of 15:1, 1 h reaction time at 160 °C reaction temperature and agitation speed of 600 rpm [29].

Similarly, nano-catalysts are also becoming very interesting for the production of biodiesel from low-quality feedstock as they do have higher catalytic activity due to having large pore size and large surface area. Having large pore size and large surface area means possessing a more active catalytic surface, because the active surface of a catalyst, which is its vital property, increases when the size of the catalyst is reduced [30]. Such higher catalytic character enables the use of a smaller amount of the catalyst compared to other catalyst options, and this has considerable economic benefits for large-scale production processes. Generally, by using nano-catalysts, better conversion of oil feedstock to biodiesel can be achieved at relatively medium temperature and shorter reaction time. The study done by Bet-Moushoul et al. [31] indicated that the oil conversion range of 90%–97% could be attained within 3 h using CaO-based gold nanoparticles as a heterogeneous catalyst for transesterification of sunflower oil with methanol. For this conversion, the optimum reaction conditions were a reaction temperature of 65 °C, methanol to oil molar ratio of 9:1, and a catalyst loading of 3 wt.% [31]. Table 1 shows some of the recent studies done on the optimum reaction conditions required to produce biodiesel from different feedstock types using bulk CaO, ionic liquid, enzyme, and nanoparticle catalysts.

Table 1. Optimum reaction conditions for biodiesel production from different feedstock using four different catalyst categories: bulk CaO, enzyme, ionic liquid, and nanoparticle catalysts.

Catalyst	Feedstock	Alcohol	Optimum Reaction Conditions				Conversion (%)	Time (h)	Ref.
			Alcohol Molar Ratio	Catalyst Amount (Wt.%)	Temperature (°C)				
Glycerol-enriched CaO	Jatropha oil	Methanol	9:1	15	65		93.5	7	[13]
CaO	Vegetable oil	Methanol	6:1	3	65		100	1.25	[32]
Activated CaO	Sunflower oil	Methanol	13:1	3	60		94	1.67	[33]
CaO from river snail shell	WCO [f]	Methanol	9:1	3	65		92.5	1	[22]
CaO from chicken manure	WCO	Methanol	12:1	7.5	65		93	6.5	[34]
CaO from WBCS [e]	WCO	Methanol	12:1	7	65		94.25	1	[35]
Novozyme 435	Soybean oil	Ethanol	3:1	5	37		95	24	[25]
Mixed Enzyme [a]	Used Palm oil	Ethanol	3:1	10	45		89	12	[26]
Immobilized lipase	Canola oil	Methanol	6:1	2.15	40		90	24	[36]
Pseudomonas fluorescens enzyme	Waste frying oil	Methanol	3:1	5	45		63	24	[37]
[Bmim]Im [b]	Vegetable oil	Methanol	6:1	6	60		95	1	[38]
4B [c]	Vegetable oil	Ethanol	9:1	5	60		94.3	5	[39]
[Hnmm]OH [d]	Soybean oil	Methanol	8:1	4	70		97	1.5	[40]
[CyN1,1PrSO3H][p-TSA]	Palm oil	Methanol	24:1	3	120		98.7	2.5	[28]
Imidazole-based I.	Tung oil	Methanol	21:1	5	120		98	2	[41]
Iron doped zinc oxide nano-catalyst	Castor oil	Methanol	12:1	14	55		91	0.833	[42]
CaO/Au nanoparticles	Sunflower oil	Methanol	9:1	3	65		90-97	3	[31]
Functionalized CaO nanoparticles	Canola oil	Methanol	9:1	3	65		97	8	[43]
KF/CaO catalyst	Tallow seed oil	Methanol	12:1	4	65		97.5	2.5	[44]

[a] Mixture of *Pseudomonas fluorescens* and *Candida rugose*; [b] Ionic liquid: 1-butyl-3-methylimidazolium imidazolide; [c] Ionic liquid: 1-benzyl-1H-benzimidazole; [d] Ionic liquid: 1-butyl-3-methyl morpholine hydroxide; [e] WBCS - white bivalve clamshell; [f] WCO - waste cooking oil.

Even though these four catalyst categories are technically capable of producing fuel quality biodiesel from various feedstock options, the relative economic feasibility of each production alternative remains unclear, as there are no such considerable studies performed so far to investigate the economic competitiveness of the alternatives. Accordingly, this study aimed at evaluating the techno-economic performances of bulk CaO, enzyme, ionic liquid, and nanoparticle catalyst technologies to produce biodiesel fuel from low-quality and cheap oil feedstock. The study would compare the proposed catalyst technologies in terms of their technological efficiencies and economic feasibility. Such an approach would give a complete view of the practicability of the process routes for sustainable production of biodiesel fuel. In addition, the study could be used as a preliminary estimate of the whole set up of the projects based on which detail estimates for the actual implementation of the efficient and affordable production process could be carried out. As to our knowledge, there are no similar investigations and comparisons performed among the catalyst technologies mentioned in this study.

The study was entirely based on process simulation involving all the unit procedures required to produce fuel quality biodiesel. These process simulations were designed using two commercial software - Aspen Plus and SuperPro. The technical performance evaluation was done based on the relative amount and purity of the product biodiesel and the byproduct glycerol, as well as the relative amount of biodiesel produced per amount of oil feedstock. Whereas the economic performance assessment was performed using economic parameters, such as total investment cost, unit production cost, net present value (NPV), internal rate of return (IRR), return over investment (ROI), and gross margin. The sensitivity of the technology options towards the change in market values of oil purchasing cost, catalyst purchasing cost, and biodiesel price was also assessed using NPV as a parameter.

2. Materials and Methods

2.1. Description of Raw Materials

The raw materials used in all the technological options include acidic oil feedstock, ethanol, and four catalyst types, such as bulk CaO, enzyme (Novozyme 435), ionic liquid (1-benzyl-1H-benzimidazole-based IL), and nano-CaO (zinc doped CaO nanoparticle). We took acidic oil with 10% FFA content to represent the oil from most of the non-edible plants [45,46], which are cheap and found everywhere. The alcohol considered was ethanol because it is non-toxic (thus easy to handle) and can be produced from renewable resources, making the biodiesel produced to be entirely from renewable resources.

The four catalyst categories considered in this study are proved to achieve the significant conversion of low-quality oil to biodiesel [25,39,47,48]. The bulk CaO-based catalyst can be prepared using cheap resources through very simple process steps like calcination [22]. Thus, we considered this catalyst because it is very cheap and can be easily prepared from waste materials. It can also be reused 13 times [49], favoring a considerable reduction of the total cost required for catalyst purchase. Concerning the nanoparticle catalyst category, we took a zinc doped nano-CaO catalyst because it does have additionally better catalytic activity due to its higher surface area [30,50]. Kumar et al. [48] found out that zinc doped CaO nano-catalyst could catalyze transesterification of oil with 8.4 wt% FFA content for its complete conversion. The third catalyst category considered was an enzyme, which is well known for its technical efficiency in producing fuel quality biodiesel from feedstock with very high FFA content [51,52]. In this study, we considered the commercial enzyme, Novozyme 435, produced from *Candida antarctica*. Li Deng et al. [53] studied the performances of different lipases with different alcohols to produce biodiesel from sunflower oil and found out that Novozyme 435 is preferable enzyme catalyst for the highest yield of fatty acid alkyl esters (with more than 90% yield) using methanol, ethanol, 1-propanol. Even though the cost of the enzyme is very high, the greater reusability rate of such catalysts would reduce the total cost required to purchase the enzyme catalyst. According to Andrade et al. [54], immobilized enzymes like Novozyme 435 could be reused 300 times, favoring the reduction of the total cost. The fourth catalyst considered was an ionic liquid catalyst, which is very well known to achieve higher conversion within relatively short reaction time when compared to most of the catalyst types used for biodiesel production [27,29,41]. In this specific study, we preferred to take the Bronsted acid ionic liquid, *1-benzyl-1H-benzimidazole,* because this catalyst is proved to be one of the highly efficient catalysts compared to other ionic liquid catalysts [39]. This catalyst can be reused 8 times without a considerable reduction in its catalytic activity [39].

2.2. Design Assumptions

The process flow diagrams of all the production technology options were designed based on the following assumptions:

- The feeding rate of the oil feedstock was kept the same for all technological options, and it was 5177.23 kg/h. This value was assumed to represent large-scale production capacity considering that the oil feedstock has 10% FFA on a molar basis; in that case, the feedstock consists of 5000 kg triglyceride and 177.23 kg FFA.
- It was assumed that there is no solid particle in the oil feedstock.
- The oil supply is continuous throughout the year.
- 7920 working hours or 330 working days per year were considered.
- In all of the equipment, the pressure drop was neglected.
- The triglyceride was represented by triolein with a density of 907.8 kg/m^3, the FFA was denoted by oleic acid with a density of 895 kg/m^3, and the pure biodiesel was denoted by ethyl oleate with a density of 873.9 kg/m^3.
- Due to the presence of polar compounds, such as ethanol and glycerol, in all of the processes considered, the non-random two liquid (NRTL) thermodynamic model was selected as the property package for the calculation of activity coefficient of the liquid phase in the simulations.
- The total project lifetime was assumed to be 15 years.
- There was no loan considered for all the projects.
- In each process option, the reusability of the catalysts was considered in the calculation of the total cost of catalyst.

2.3. Description of the Technology Options for Biodiesel Production

Four different catalyst options for biodiesel production from acid oil were considered in order to examine their techno-economic performances while producing fuel quality biodiesel from cheap oil. *Technology option I*: The enzyme-catalyzed transesterification and esterification; *Technology option II*: Bulk CaO-catalyzed transesterification; *Technology option III*: Ionic liquid-catalyzed transesterification; and *Technology option IV*: Nano-CaO-catalyzed transesterification. Recently, these catalyst technology options are getting more emphasis by researchers for efficient and eco-friendly production of biodiesel from cheap resources. In all the process alternatives, transesterification is the dominant reaction; however, other possible side reactions may occur based on the oil quality and the type of the catalyst used. The dominant reactions, the optimum reaction conditions, the amount and specific type of input materials, as well as the whole flow of the processes involved in each catalyst technology, are indicated as follows.

2.3.1. Technology Option I

This option was designed to investigate the techno-economic performance of the enzyme-catalyzed biodiesel production process by involving all the equipment necessary to get fuel quality biodiesel. Figure 1 indicates the process flow diagram of the enzyme catalysis technology option.

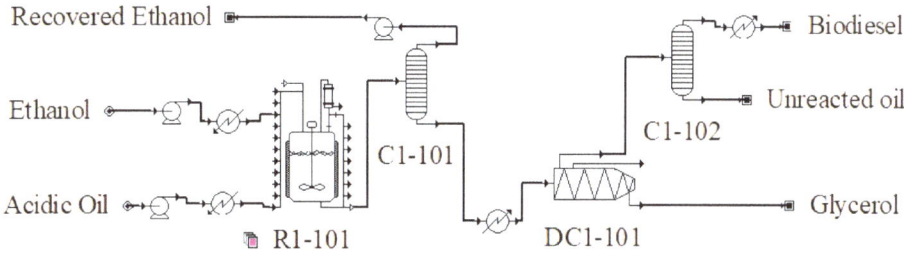

Time Ref: h		Acidic Oil	Ethanol	Biodiesel	Glycerol	Recovered Ethanol	Unreacted oil
Type		Raw Material	Raw Material	Revenue	Revenue	Credit	Credit
Total Mass Flow	kg	5177.23	809.20	5197.04	507.40	39.54	242.54
Temperature	°C	25.0	25.0	25.0	38.5	78.3	359.2
Pressure	bar	1.013	1.013	1.013	1.013	2.013	1.013

Figure 1. Technology option I: R1-101 continuous stirred-tank reactor (CSTR) to produce biodiesel, C1-101 first short cut distillation column to recover unreacted ethanol, DC1-101 centrifugal decanter to separate glycerol, and C1-102 second short cut distillation column to purify biodiesel.

The dominant reactions involved in enzyme-catalyzed processes are transesterification and esterification. There was also hydrolysis of the triglyceride by water produced from the esterification reaction. Therefore, enzyme-catalyzed biodiesel production was comprised of two processes, namely: direct alcoholysis of triacylglyceride in one-step reaction and two-step hydrolysis of triacylglyceride followed by esterification [25].

The optimum reaction condition was taken to be 3:1 ethanol to oil molar ratio, 5 wt.% Novozyme 435 catalyst, and 37 °C reaction temperature to attain about 95% oil conversion within 24 h [25]. The alcohol to oil molar ratio considered here was the exact stoichiometric amount (3:1) because an excessive amount of alcohol in the reaction could hinder the activity of the enzyme [25,55]. Especially when methanol is used as the reacting alcohol, the effect is more pronounced [25], and it is always recommended to perform stepwise (2 or 3 steps) addition of the alcohol to the reaction [51,56]. However, concerning ethanol alcohol, the effect is not that significant, and the one-step addition of the stoichiometric amount does not significantly affect the enzyme activity. This might be due to the lower amount of undissolved alcohol in the substrate when we use ethanol than methanol because it is much amount of undissolved alcohol that inhibits the enzyme activity [57]. Thus, since ethanol is more soluble in oil than methanol [58], enzyme inhibition effect is very low when we use ethanol than methanol. Cervero' et al. [25] also indicated that at maximum reaction time, the conversion of soybean oil to biodiesel was almost similar for both single step and multiple step addition of ethanol to the reaction. Accordingly, a one-step addition of the ethanol was considered in this process flow. This could also avoid the need to include more reactors, which would otherwise be if the alcohol is added in multiple steps.

Both the oil (5177.23 kg/h) and ethanol (809.35 kg/h) were heated up to 37 °C and pumped separately to a continuous stirred-tank reactor (R1-101), which has a total volume of 33.6 m^3 and packed with Novozyme 435 catalyst. The reactor was designed to have a constant temperature of 37 °C and work continuously in such a way that the oil conversion of 95% could be achieved within a residence time of 24 h based on the optimum reaction conditions taken from the literature [25]. The produce from the reactor was then directed to the first distillation column (C1-101) to recover the unreacted ethanol for possible reuse and to improve the biodiesel quality too. The bottom outlet from this distillation column was cooled down and directed into a centrifugal decanter (DC1-101) to separate the glycerol. The upper output from this centrifugal decanter was then taken to the second short cut distillation column (C1-102) to purify the biodiesel. This distillation column was designed to have 11 number of stages and 0.125-reflux ratio for which the maximum possible biodiesel purification could be attained.

2.3.2. Technology Option II

The second technological option considered the application of bulk CaO catalyst to produce biodiesel through transesterification of acidic oil by involving all unit procedures required to get high-quality fuel. Figure 2 indicates the flow diagram of the whole processes involved to produce biodiesel fuel using the CaO catalyst.

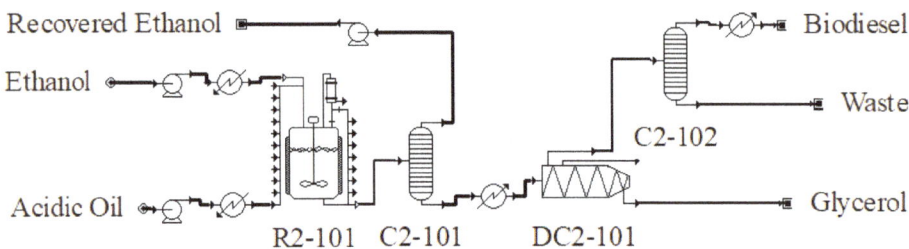

Time Ref: h		Acidic Oil	Ethanol	Biodiesel	Glycerol	Recovered Ethanol	Waste
Type		Raw Material	Raw Material	Revenue	Revenue	Credit	
Total Mass Flow	kg	5177.2300	2341.3500	5132.4253	503.2338	1624.0040	276.5980
Temperature	°C	25.0	25.0	25.0	26.7	52.5	259.0
Pressure	bar	1.013	1.013	0.150	1.013	3.750	0.150

Figure 2. Technology option II: R2-101 CSTR for the production of biodiesel, C2-101 first short cut distillation column to recover excess ethanol, DC2-101 centrifugal decanter to separate the glycerol, and C2-102 second short cut distillation column to purify the biodiesel.

In this technology option, the dominant reaction was a CaO-catalyzed transesterification reaction for which excess amount of ethanol was used to favor forward reaction for more biodiesel production [20,32]. There was also an unavoidable saponification reaction between the FFA and the catalyst, which could not be dominant due to the relatively lower amount of FFA. The reactor designed was a continuous stirred-tank reactor packed with a bulk CaO catalyst. The optimum reaction conditions taken into consideration were oil to ethanol molar ratio of 9:1, catalyst loading of 7 wt.% with respect to oil, and reaction temperature of 75 °C; and at such reaction conditions, 97.58% oil conversion could be achieved within 2 h [47].

Oil amount at 5117.23 kg/h and the ethanol amount at 2341.35 kg/h were heated up to 75 °C separately and pumped into the continuous stirred-tank reactor (R2-101), which has a total volume of 18.9 m^3 and packed with bulk CaO catalyst. The reactor was designed to have a 75 °C constant temperature. The outlet from the reactor was directed to the first distillation column (C2-101) to separate the excess ethanol for reuse. Seven stages and 2-reflux ratio were the optimum values taken in the design of this distillation column to recover the maximum possible ethanol left after the reaction. The lower pipe from this distillation column was directed to a centrifugal decanter (DC2-101) for glycerol separation from the product mixture. The upper outlet from this centrifugal decanter was then directed to the second distillation column (C2-102) for purification of the biodiesel product. This distillation column was designed with 4 number of stages and 3 reflux ratios, beyond which there could not be further purity of the biodiesel attained. The waste stream from this process was composed of unconverted oil and calcium soap, which is non-toxic and rather useful if further purification is included. However, such an additional purification unit procedure incurs the considerable cost and would increase the overall production cost, making the technology option economically unattractive.

2.3.3. Technology Option III

In this technology option, the ionic liquid-catalyzed biodiesel production process was designed for techno-economic evaluation of the possible arrangement of all equipment required to produce fuel quality biodiesel. Figure 3 indicates the whole flow diagram required to produce fuel quality biodiesel using a specific type of ionic liquid catalyst.

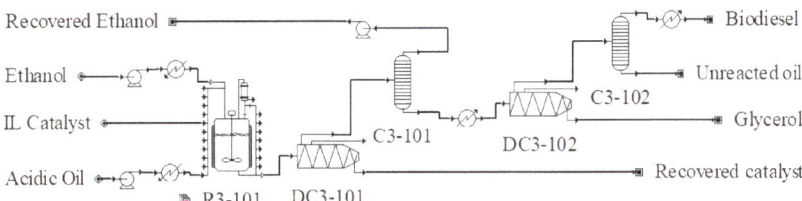

Time Ref: h		Acidic Oil	Ethanol	Ionic Liquid Catalyst	Biodiesel	Glycerol	Recovered ethanol	Recovered Catalyst	Unreacted oil
Type		Raw Material	Raw Material	Raw Material	Revenue	Revenue	Credit	Credit	Credit
Total Mass Flow	kg	5177.2300	2428.0700	89.0000	5104.1441	490.1182	1674.3256	89.0761	336.7259
Temperature	°C	25.0	25.0	25.0	25.0	26.7	78.3	60.0	101.0
Pressure	bar	1.013	1.013	1.013	1.013	1.013	4.513	1.000	1.013

Figure 3. Technology option III: R3-101 CSTR for the production of biodiesel, DC3-101 first centrifugal decanter to recover the catalyst, C3-101 first distillation column to recover excess ethanol, DC3-102 second centrifugal decanter to separate the glycerol, and C3-102 second short cut distillation to purify the biodiesel.

Transesterification was the dominant reaction considered here, even though there was also an esterification reaction due to the presence of FFA in the oil. The optimum reaction condition taken into consideration for this process option was 9:1 ethanol to oil molar ratio, 5% (based on mmol of oil) catalyst, and 60 °C reaction temperature to attain a maximum conversion (94.3%) of the oil within 5 h [39].

Oil with a rate of 5177.23 kg/h and ethanol with a rate of 2428.07 kg/h were heated up to 60 °C separately and pumped into CSTR (R3-101), which has a total volume of 23.7 m^3 and to which a Brønsted acid ionic liquid (1-benzyl-1H-benzimidazole) catalyst was also supplied at a rate of 258.86 kg/h. The reactor was designed to work at a constant temperature of 60 °C. The product from this reactor was directed into the first centrifugal decanter (DC3-101) for the separation of the catalyst from the remaining product mixture. The upper outlet from this centrifugal decanter was let into the first short cut distillation column (C3-101) to recover the leftover ethanol for recycling. This column was designed to have 5 number of stages and 3.5-reflux ratio, above which there was no change in amount and quality of ethanol recovered. The bottom output from this first distillation column was then directed to the second centrifugal decanter (DC3-102) to separate glycerol. The upper outlet from the second centrifugal decanter was let into the second distillation column (C3-102) for the purification of the biodiesel product. This distillation column was designed to have 7 actual stages and 0.125 reflux ratio by which the maximum possible purity could be attained.

2.3.4. Technology Option IV

As the fourth technology option, the nano-CaO-catalyzed process was designed to assess the techno-economic performance for the production of fuel quality biodiesel. Figure 4 indicates the whole process flow diagram of producing fuel quality biodiesel using zinc doped CaO nano-catalyst.

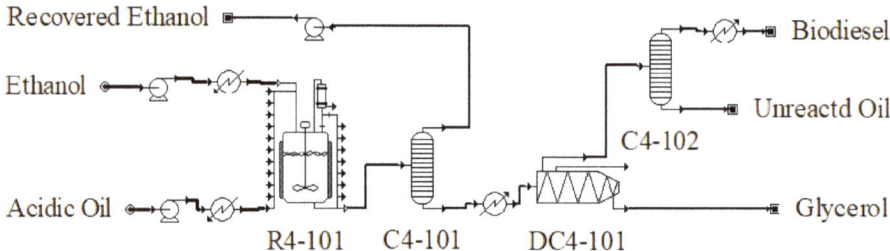

Time Ref: h		Acidic Oil	Ethanol	Biodiesel	Recovered Ethanol	Unreactd Oil	Glycerol
Type		Raw Material	Raw Material	Revenue	Credit	Credit	Revenue
Total Mass Flow	kg	5177.2300	2341.3500	5202.9529	1569.8496	232.2409	513.6241
Temperature	°C	25.0	25.0	25.0	78.3	353.1	26.5
Pressure	bar	1.013	1.013	1.013	4.513	1.013	1.013

Figure 4. Technology option IV: R4-101 CSTR for the production of biodiesel, C4-101 first short cut distillation column to recover excess ethanol, DC4-101 centrifugal decanter to separate the glycerol, and C4-102 second distillation column to purify the biodiesel.

Transesterification was the dominant reaction using the nano-CaO catalyst. According to the study done by Kumar et al. [48], the existence of a saponification reaction among the specific catalyst, zinc doped CaO, and the FFA in the oil is negligible. The optimum reaction condition for 99% conversion of the oil within 1 h was taken to be the molar ratio of ethanol to the oil of 9:1, catalyst amount of 5 wt.% with respect to oil, and 65 °C as the reaction temperature [48].

The oil at a feeding rate of 5177.23 kg/h and ethanol at a rate of 2341.35 kg/h were heated up to 65 °C separately and driven into the continuous stirred-tank reactor (R4-101), which has a total volume of 9.4 m^3 and packed with zinc doped CaO nano-catalyst. The rector was designed to work at a constant temperature at 65 °C. The produce coming out of the reactor was directed into the first distillation column (C4-101) to distill out the excess ethanol for reusing. This column was designed to have 4 number of stages and 1 reflux ratio for the maximum possible recovery of the excess ethanol. The bottom outlet from the first distillation column was cooled down to ambient temperature and directed to a centrifugal decanter (DC4-101) for the separation of the glycerol from the rest of the mixture. Finally, the upper outlet from this centrifugal decanter was directed into the second short cut distillation column (C4-102) to purify the biodiesel from impurities, such as unreacted oil and remaining glycerol. This distillation column was designed to have 7 stages and 0.2 reflux-ratio at which the maximum possible purity of the biodiesel product could be achieved.

In all the production technology options, the storage tanks for raw materials and output were not involved because the raw materials are considered to be used immediately, and the outputs could also be used as soon as they are produced without the need to store them. In most of the process options, there was no waste stream from the production, except in technology option II, where the waste stream was composed of unreacted oil and calcium soap. This waste stream could be purified further to get reusable oil and economically valuable calcium soap. Calcium soap is vital as fat supplements for ruminants because it comprises a high concentration of fat and calcium, and both are beneficial for ruminants [59].

2.4. Techno-Economic Assessment

The technical performances of the technology options were evaluated based on the relative amount and purity of biodiesel product and glycerol byproduct while using the same amount and quality of oil feedstock. The other important parameter considered was the quantity of biodiesel that could be produced from a kilogram of the oil feedstock. Such technical performance assessment depends on the material and energy balance, which was done using Aspen plus V10 considering optimum reaction conditions of the dominant reactions in each technology option.

The economic analysis of the processes was carried out using SuperPro software. By using financial input data, the program calculated the internal return rate (IRR) (before and after-tax), NPV (at 7% interest rate), gross margin, unit production cost, and annual revenue, among other important economic parameters. The latest market values and the estimated cost of raw materials, utilities, labor, and equipment were used as the basis for evaluating the economic performances of the studied technology options. The other considerable cost categories for such evaluation were equipment installation cost, auxiliary facilities cost, and depreciation cost, among others. The feedstock taken was non-edible and cheap oil with an estimated cost in the range of 478–684 US$/ton [45,46,60]. In this specific study, feedstock cost of 580 US$/kg was taken as the average value because it is very cheap to produce such oil in the Ethiopian context, even though there is no formal market to buy or sell non-edible oil in the country. The delivered cost of the other raw materials, such as the four catalysts and ethanol, were based on the latest market prices taken from various sellers in Ethiopia, as well as from the relevant literature [54,61].

The costs of all the required labor categories are according to the current wage indicator in Ethiopia [62], for which the conversion to US$ was done based on the rate at the time of referring the database. The labor cost was calculated using the basic rates allocated for each labor category. In doing so, the basic rate was multiplied by the sum of the benefit, supervision, supplies, and administration rates, as well as the total labor hours. In all the technology options, the percent of work time dedicated to process-related activities, which is used to estimate the labor time, was taken to be 70%, considering that the technology options involve continuous processes. The utilities considered in all the technology options include electricity, steam, steam high, and cooling water, and their cost estimations were taken according to the current market prices in Ethiopia as well as from the relevant literature [63,64]. Table 2 shows the estimated costs for utilities, labor, and raw materials considered in all the technology options.

Table 2. The estimated cost of raw materials, labor, and utilities used in the four technology options.

Raw Materials	Cost
Oil	0.478 US$/kg
Ethanol	0.300 US$/kg
Bulk CaO	0.120 US$/kg
Ionic Liquid	50.5 US$/kg
Enzyme (Novozyme 435)	1000 US$/kg
Nano-CaO	6.5 US$/kg
Utilities	
Electricity	0.021 US$/KW-h
Steam high	10 US$/MT
Steam	6 US$/MT
Cooling water	0.025 US$/MT
Labor (Basic rate)	
Reactor operator	15 US$/h
Operator	10 US$/h

The cost of every equipment involved in all technology options was estimated using the Peter and Timmerhaus method [65]. For such estimation, the Chemical Engineering Plant Cost Index of 691.8 for February 2019 was used [66]. This index signifies the money time value due to deflation and inflation by which the average cost of each equipment can easily be calculated for the year 2019 using previous year cost values. For the estimation of the other components of the capital investment cost like instrumentation, piping, electricity, installation, and yard improvement, a method involving allocation of a percentage of total equipment purchasing cost was used based on literature, as shown in Table 3.

Table 3. Direct plant cost categories and their percentage allocation with equipment cost [67].

Cost Category	% Allocation with Equipment
Piping	20
Instrumentation	10
Electrical	15
Insulation	3
Building	15
Yard improvement	10
Auxiliary facilities	25
Unlisted equipment	20

These capital investment cost categories could directly be used in SuperPro because its cost estimation interface gives options to assign an estimated percent of total equipment cost for each direct plant cost category. The other equipment-associated costs, such as insurance, depreciation, maintenance cost, and tax, could also be put in the software based on the percentage allocation of their costs, as indicated in Table 4.

Table 4. Cost estimation methods for components of capital investment and operating costs.

Cost Items	Estimation Method
Capital investment cost categories	
Installation cost (for each equipment)	$0.2 \times PC$ [h]
Maintenance cost (for each equipment)	$0.1 \times PC$
Purchasing cost of unlisted equipment (PCUE)	$0.2 \times PC$
Installation cost of unlisted equipment	$0.5 \times PCUE$
Operating cost categories	
Insurance	$2 \times DFC$ [i]
Local tax	$15 \times DFC$
Factory expense	$5 \times DFC$
Laboratory and quality control	$30 \times TLC$ [j]

[i] DFC–direct fixed cost; [h] PC–equipment purchasing cost; [j] TLC–total labor cost.

3. Results and Discussion

The material and energy balance of the four technology options were carried out based on determined equipment size and the optimum reaction conditions taken for each dominant reaction in the processes. Using the results from material and energy balance together with the latest prices of raw materials, utilities, labor, and equipment, the techno-economic performances of the technology options have been evaluated and presented as follows.

3.1. Technical Performances

All the process options could provide fuel quality biodiesel and pure glycerol, proving that the catalysts used together with the unit procedures involved in separation and purification of the crude biodiesel could attain high-quality products. Accordingly, the biodiesel from all technology options fulfilled the American Society for Testing and Materials (ASTM) standards for biodiesel fuel quality. However, there was a slight variation in the amount of biodiesel and glycerol produced. In terms of biodiesel product, technology option IV had the highest performance, with about 98.98 kg/h product variation from the least performing one. This was mainly owing to the high catalytic activity of the nano-CaO particles, which favors the high conversion of the oil into biodiesel within relatively short reaction time. It might also be due to the negligible occurrence of the saponification reaction when zinc doped nano-CaO catalyst was used [48], which also minimizes the likeliness of the catalyst being used by the FFA in the process of saponification. Relatively, the least performance in terms of biodiesel product was indicated in the ionic liquid catalyst option. This was due to the lower conversion percentage achieved in the given optimum reaction conditions taken from the literature [39].

Similarly, the higher glycerol production amount was attained in technology option IV, with a product variation of 23.43 kg/h glycerol from the least performing option. This was again due to the variation in the achievement of oil conversion percentage according to the required optimum reaction conditions. Consequently, the option I and IV did have a relatively highest performance as they provided more amount of biodiesel from the same amount of feedstock used. Table 5 indicates the relative technical performances of the technology options studied.

Table 5. Summary of technical performances of the technology options.

Indicators	Technology Options			
	Option I	Option II	Option III	Option IV
Biodiesel amount (kg/h)	5191.26	5132.16	5103.64	5202..62
Biodiesel quality (% mass)	99.9	99.9	99.9	99.9
Glycerol amount (kg/h)	507.47	503.06	489.98	513.41
Glycerol quality (% mass)	99	99	99	99
Performances (biodiesel/oil)	1	0.991	0.986	1
Impurities in biodiesel *				
Glycerol (% mass)	0.11	0	0.01	0
Triolein (% mass)	0	0	0	0

* The maximum allowable amount of impurities, according to ASTM (American Society for Testing and Materials,) are Glycerol 0.25% mass and Triolein 0.20% mass.

The catalysts from all the technology options could be recovered and reused for a number of times. This would help to reduce a considerable amount of money, which otherwise could be spent to purchase the extra catalyst. Another advantage of the processes was that in all the process options, except technology option II, there was no waste produced. In option II, the waste stream was composed of unreacted oil and calcium soap that could be further purified for economic benefits. The unreacted oil from technology option I and option III could be recycled directly to the processes, whereas the one from technology option IV should pass through a treatment step before it is reused in order to reduce the FFA content. This is because 76% of the unreacted oil from this process was composed of FFA, which was left unreacted in the nano-catalysis process.

3.2. Economic Performances

Technology option I was the most expensive alternative, mainly due to the very high cost of the enzyme, Novozyme 435. Even though this catalyst could be repeatedly used for more than 200 times [54], and the process could give the second higher biodiesel product, the higher total investment cost of the option could not make it economically feasible for the production of biodiesel fuel. The higher production cost in option I was also attributed to its relatively larger reactor volume required due to longer reaction time. Because the larger the equipment volume, the higher would be the costs of equipment, facilities, and utilities.

The second expensive option was the technology option III. Its total investment cost was almost half of that of the option I and 37% higher than the least cost option, which was option II. This was mainly because of the second larger volume of reactor required due to longer reaction time as well as because of additional centrifugal decanter required to separate the catalyst. The larger and the more equipment we use, the higher would be the utility cost and the other equipment-associated costs. Technology option II had the least total capital investment cost because it required smaller equipment sizes due to minimum reaction time, and the catalyst involved was the cheapest among the catalyst options studied.

Even though technology option IV was the second cheapest option, the higher cost of the nano-CaO catalyst could still make it economically infeasible at the optimum market prices of raw materials and outputs. Similarly, option III was also found to be economically infeasible at the current market prices of raw materials and outputs. However, for the optimum market values of inputs and outputs considered, option II was the most feasible option with positive NPV, lower unit production cost, higher IRR, ROI, and gross margin. Table 6 summarizes the economic performances of the technology options. It highlights the comparative economic performances of the process options for the given market scenario. The first part indicates the total investment cost, followed by expenditures in cost categories. The calculated revenues from the product and byproduct, as well as the value of the calculated economic parameters, are also indicated in the Table.

Table 6. Summary of the economic performance of the technology options.

Economic Performance Parameters	Catalyst Technology Options			
	Option I	Option II	Option III	Option IV
Total capital investment cost (US$)	13,200,448	4,608,642	6,319,464	4,744,425
Total equipment purchasing cost (US$)	1,629,303	432,295	674,025	403,033
Direct fixed capital (US$)	6,716,375	1,781,747	2,778,061	1,682,115
Working capital (US$)	6,148,254	2,737,807	3,402,499	2,978,204
Total annual operating cost (US$)	71,304,387	31,224,324	39,050,943	33,824,494
Total annual raw material cost (US$)	66,706,623	29,372,952	36,670,506	32,010,394
Labor dependent cost (US$)	624,549	364,320	390,343	364,320
Facility dependent cost (US$)	3,376,225	889,148	1,396,347	844,949
Laboratory, quality control, and analysis (US$)	187,365	109,296	117,103	109,296
Utility cost (US$)	299,626	378,608	366,644	385,535
Annual revenue from Biodiesel (US$/year)	32,087,761	31,704,748	31,529,636	32,140,209
Annual revenue from Glycerol (US$/year)	1,607,422	1,594,245	1,552,758	1,627,161
Total annual revenue (US$/year)	33,695,184	33,298,993	33,082,394	33,767,370
Unit production revenue (US$/kg)	0.8186	0.8192	0.8184	0.8194
Unit production cost (US$/kg biodiesel)	1.7323	0.7681	0.9660	0.8208
Net Present Value at 7% (US$)	−349,847,116	9,736,266	−57,834,235	−3,217,935
Return over Investment (%)	26.11	84.66	71.48	103.69
After tax Internal Rate of Return (%)	−100	32.73	−100	−100
Gross margin (%)	13.79	17.53	20.19	21.93

The higher amount of biodiesel product and glycerol byproduct for the nano-CaO-catalyzed option results in a relatively higher value of total annual revenue incurred, as shown in Table 6. The lowest total annual revenue was recorded for ionic liquid-catalyzed option with about 684,976 US$/year lower than the revenue from the nano-catalyzed option. The enzyme-catalyzed option scored the highest unit production cost with about 0.964 US$ increment per kilogram of biodiesel product compared to the bulk CaO-catalyzed option. Positive after-tax IRR was recorded only for bulk CaO-catalyzed option. For the enzyme-catalyzed option, the gross margin, ROI, and NPV were the lowest, followed by the ionic liquid-catalyzed option.

Except for the bulk CaO-catalyzed option, the other three process options showed negative NPV. This indicated that the investment of each project was not profitable because the present value of the net cash flow in each project, within the projects' lifetime, was lower than the present initial cash required to establish them. This was exhibited more by the values of IRR for each process option. The after-tax IRR of the bulk CaO-catalyzed option was positive, whereas, for the other three options, the calculated amount was negative, meaning that the projects did not perform well over time. For instance, without considering the discounted cash flow, i.e., only based on the amount of return and cost of investment, the nano-CaO-catalyzed option seemed efficient as it had a higher percentage of ROI. However, its NPV and IRR were negative, implying that the investment in this process option is not viable within the given lifetime.

Concerning the relative economic performances of the technologies, divergent results might be obtained if calculations are done in a different market scenario or using market values of inputs and outputs, which are not comparable to what has been used in this study. This implies that such performances are expected to be different for countries with different market scenarios.

3.3. Sensitivity Analysis

The economic feasibility of the studied technology options was very diverse, mainly due to the cost variation among the catalysts as well as the number and size of equipment required to attain fuel quality biodiesel. Thus, it seems reasonable to test how sensitive the technology options are towards the change in market values of the inputs and outputs. Among the various economic variables, oil cost and catalyst cost comprise the higher percentage of the raw materials' cost. Similarly, biodiesel is the main product to get the desired revenue from the projects. Therefore, in this study, the economic effect of variations of oil purchasing cost, biodiesel selling price, and catalyst purchasing cost was evaluated in terms of NPV (at 7% interest rate); and the results among the technology options were compared and presented as follows. We considered biodiesel price since biodiesel is the main product, and its price fluctuation could have a direct effect on the feasibility of the businesses. We considered oil cost because the cost of feedstock took a higher share of raw material cost. In addition, we took the cost of catalysts for sensitivity analysis in order to indicate how the respective cost of the studied catalysts affect the businesses as well as to indicate the maximum possible cost of each catalyst for the economic feasibility of the businesses.

3.3.1. Effect of Change of Oil Cost on NPV

The trend at which the technology options respond towards change in oil purchasing cost was almost similar. However, the option I was found to be economically infeasible for all ranges of the oil purchasing cost considered. For option III, the maximum cost of oil feedstock was about 0.39 US$/kg, beyond which the option would be economically infeasible. Option II was found to be more tolerant of the market variation of oil cost. It could, still, be economically feasible up to 0.59 US$/kg of the oil purchasing cost. In comparison, option IV was found to be the second most tolerant of market fluctuations of the oil purchasing cost. Nevertheless, it could be economically feasible for oil purchasing cost less than 0.51 US$/kg. Figure 5 indicates the effect of the change in oil purchasing cost on NPV of the technology options.

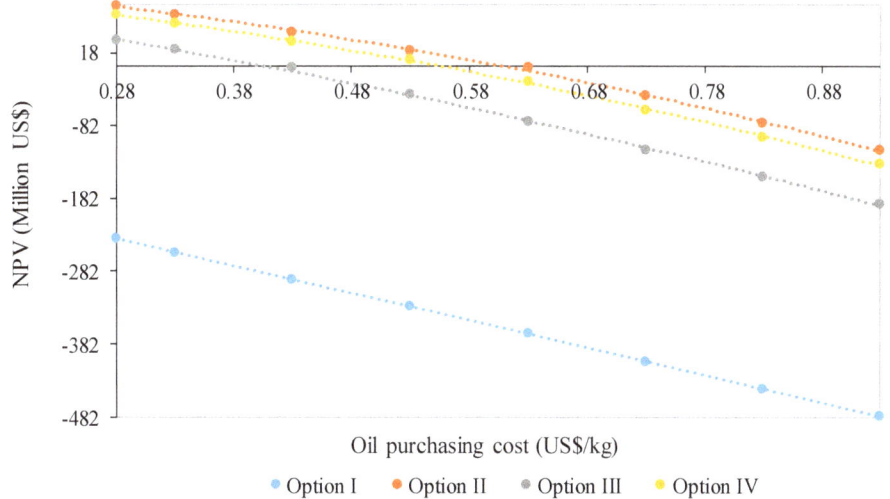

Figure 5. Effect of change of oil purchasing cost on NPV (net present value).

3.3.2. Effect of Change of Biodiesel Selling Price on NPV

The changing trend of the technology options towards the variation of the market values of biodiesel price was almost similar. Option I was unprofitable for the considered ranges of biodiesel selling prices. For this option to be economically feasible, biodiesel should be sold at a very high price (1.8 US$/kg), which is practically impossible. On the contrary, technology option IV could be economically feasible, with almost half of this price (0.97 US$/kg). This designates that the production of biodiesel fuel using technology option I should be subsidized to make the fuel economically competitive with fossil diesel in the market. For the biodiesel selling price range considered, option II was found to be more tolerant of the possible fluctuation of biodiesel price and could still be economically feasible at a biodiesel price as low as 0.77 US$/kg. Figure 6 indicates the effect of variation of the biodiesel selling price on NPV of the technology options.

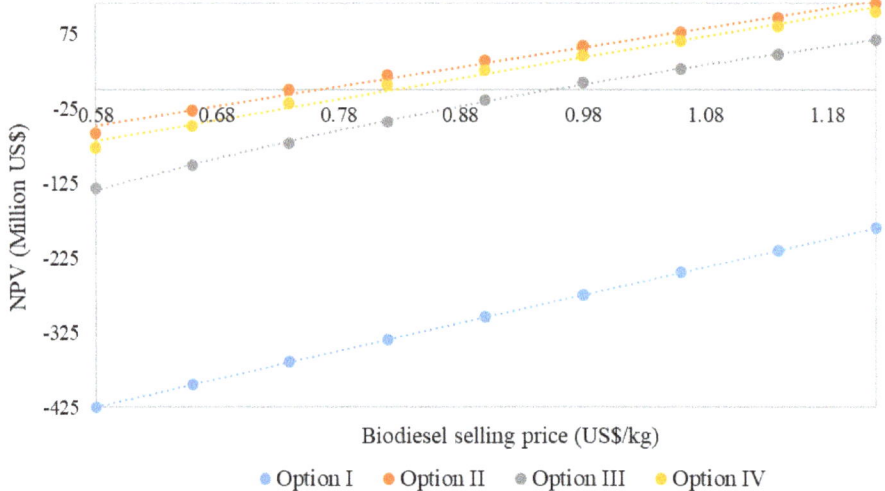

Figure 6. Effect of change of biodiesel selling price on NPV.

3.3.3. Effect of Change of Catalyst Purchasing Cost on NPV

The entire techno-economic comparison was among these four catalyst technologies. Thus, evaluating the effect of the market values of the four catalyst types could give reasonable ground for selecting the technology option that is more tolerant of fluctuating cost of materials in the market. The trend at which the NPV changes with the catalyst cost was almost the same for all technological options. However, some get negative NPV at very low-cost values and some at relatively higher values. For instance, an option I could still be economically feasible for about 61 US$/kg cost of an enzyme catalyst. This was mainly due to its higher reusability that could reduce the total cost of the catalyst. Nevertheless, this higher price, indicated here, is not enough to buy the very expensive enzyme catalysts, especially immobilized ones [54]. This demands more investigation on enzyme catalysts, which could be produced with a cost as low as 60 US$/kg while possessing the same catalytic performance as indicated here. Option III was found to be the most sensitive towards a change in the value of catalyst purchasing cost. It was economically feasible for a catalyst cost of less than 4.1 US$/kg. Option IV got its negative NPV for a catalyst cost of more than 5 US$/kg. The cheapest catalyst was the bulk CaO catalyst in option II. It could be prepared from wastes, such as eggshell, crab shell, and river snail shell, among others. The cheap cost and higher reusability of the CaO catalyst made option II more tolerant of the possible fluctuations of catalyst purchasing cost in the market. It got its negative NPV for a catalyst cost more than 7 US$/kg, which seems to be far from its current market value, as indicated in Table 2. Figure 7 shows the effect of variation of catalyst cost on the NPV of each technology option.

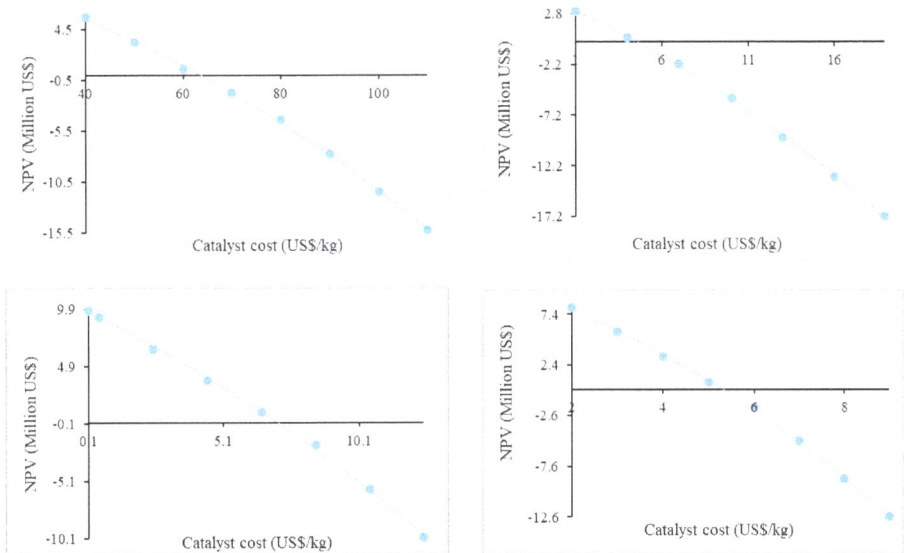

Figure 7. Effect of change of catalyst purchasing cost on NPV of each production technology option.

4. Conclusions

All the studied technology options could produce fuel quality biodiesel and pure glycerol. Their technical performance regarding the quantity of biodiesel per amount of oil feedstock was almost the same. Economically, the enzyme-catalyzed option was not feasible, mainly due to the very high cost of the enzyme catalyst and a larger volume of the reactor. The second expensive technology was the ionic liquid-catalyzed option. This is because it had the second-largest reactor volume and a greater number of equipment required to get fuel quality biodiesel. The bulk CaO-catalyzed option was the most efficient in economic terms as it attained higher positive NPV, higher IRR, higher gross margin, higher ROI, and minimum total capital investment cost. The enzyme-catalyzed option was not economically feasible for all possible ranges of biodiesel price and oil cost considered. The enzyme catalyst had to be bought for less than 60 US$/kg for the process to be economically feasible at all. The bulk CaO-catalyzed option was the most tolerant of the change in the price of biodiesel, oil cost, and catalyst cost. It indicated profitability at a biodiesel price as low as 0.74 US$/kg, oil purchasing cost as high as 0.70 US$/kg, and catalyst cost as high as 7 US$/kg.

Author Contributions: Conceptualization, S.N.G. and J.M.M.; Methodology, S.N.G. and J.M.M.; software, S.N.G. and J.M.M.; validation, S.N.G., T.H.-E., and M.T.T.; formal analysis, S.N.G. and J.M.M.; investigation, S.N.G. and J.M.M.; resources, M.T.T.; writing—original draft preparation, S.N.G. and J.M.M.; writing—review and editing, T.H.-E., M.T.T., and J.M.M.; supervision, T.H.-E., M.T.T., and J.M.M.; project administration T.H.-E. and M.T.T.; funding acquisition, T.H.-E. and M.T.T.

Funding: The financial support is from the Norad/EnPe project entitled "Research and Capacity Building in Clean and Renewable Bioenergy in Ethiopia", with Agreement No. 2015/8397 to the Norwegian University of Life Sciences (NMBU).

Acknowledgments: The financial support from the Norad/EnPe project entitled "Research and Capacity Building in Clean and Renewable Bioenergy in Ethiopia", Agreement No. 2015/8397 to the Norwegian University of Life Sciences (NMBU), is gratefully acknowledged.

Disclaimer: The authors are not responsible for any decision that might be made using results from the process options. The process options specified in this study are merely for research purposes. If anyone needs further clarification or wants to use the result for specific applications, please contact the authors for more information concerning the limitations and scope of the designs.

Conflicts of Interest: All authors declare no conflicts of interest in this paper.

References

1. IAE. World Energy Outlook 2018, Excutive Summary. International Energy Agency, France by DESK. 2018. Available online: https://www.iea.org/weo2018/ (accessed on 10 October 2019).
2. Niculescu, R.; Clenci, A.; Iorga-Siman, V. Review on the Use of Diesel-Biodiesel-Alcohol Blends in Compression Ignition Engines. *Energies* **2019**, *12*, 1194. [CrossRef]
3. Chattopadhyay, S.; Sen, R. Fuel properties, engine performance and environmental benefits of biodiesel produced by a green process. *Appl. Energy* **2013**, *105*, 319–326. [CrossRef]
4. Bajpai, D.; Tyagi, V.K. Biodiesel: Source, production, composition, properties and its benefits. *J. OLEo Sci.* **2006**, *55*, 487–502. [CrossRef]
5. Arbab, M.I.; Masjuki, H.H.; Varman, M.; Kalam, M.A.; Imtenan, S.; Sajjad, H. Fuel properties, engine performance and emission characteristic of common biodiesels as a renewable and sustainable source of fuel. *Renew. Sustain. Energy Rev.* **2013**, *22*, 133–147. [CrossRef]
6. Atadashi, I.M.; Aroua, M.K.; Abdul, A.A.R.; Sulaiman, N.M.N. The effects of catalysts in biodiesel production: A review. *J. Ind. Eng. Chem.* **2013**, *19*, 14–26. [CrossRef]
7. Leung, D.Y.; Wu, X.; Leung, M.K. A review on biodiesel production using catalyzed transesterification. *Appl. Energy* **2010**, *87*, 1083–1095. [CrossRef]
8. Yusuf, N.; Kamarudin, S.; Yaakub, Z. Overview on the current trends in biodiesel production. *Energy Convers. Manag.* **2011**, *52*, 2741–2751. [CrossRef]
9. Helwani, Z.; Aziz, N.; Bakar, M.; Mukhtar, H.; Kim, J.; Othman, M. Conversion of Jatropha curcas oil into biodiesel using re-crystallized hydrotalcite. *Energy Convers. Manag.* **2013**, *73*, 128–134. [CrossRef]
10. Gebremariam, S.N.; Marchetti, J.M. Biodiesel production technologies: Review. *AIMS Energy* **2017**, *5*, 425–457. [CrossRef]
11. Semwal, S.; Arora, A.K.; Badoni, R.P.; Tuli, D.K. Biodiesel production using heterogeneous catalysts. *Bioresour. Technol.* **2011**, *102*, 2151–2161. [CrossRef]
12. Borges, M.; Díaz, L. Recent developments on heterogeneous catalysts for biodiesel production by oil esterification and transesterification reactions: A review. *Renew. Sustain. Energy Rev.* **2012**, *16*, 2839–2849. [CrossRef]
13. Avhad, M.R.; Gangurde, L.S.; Sanchez, M.; Bouaid, A.; Aracil, J.; Martínez, M.; Marchetti, J.M. Enhancing Biodiesel Production Using Green Glycerol-Enriched Calcium Oxide Catalyst: An Optimization Study. *Catal. Lett.* **2018**, *148*, 1169–1180. [CrossRef]
14. Marwaha, A.; Rosha, P.; Mohapatra, S.K.; Mahla, S.K.; Dhir, A. Waste materials as potential catalysts for biodiesel production: Current state and future scope. *Fuel Process. Technol.* **2018**, *181*, 175–186. [CrossRef]
15. Lam, M.K.; Lee, K.T.; Mohamed, A.R. Homogeneous, heterogeneous and enzymatic catalysis for transesterification of high free fatty acid oil (waste cooking oil) to biodiesel: A review. *Biotechnol. Adv.* **2010**, *28*, 500–518. [CrossRef]
16. Marchetti, J.M.; Pedernera, M.N.; Schbib, N.S. Production of biodiesel from acid oil using sulfuric acid as catalyst: Kinetics study. *Int. J. Low-Carbon Technol.* **2010**, *6*, 38–43. [CrossRef]
17. Miao, X.; Li, R.; Yao, H. Effective acid-catalyzed transesterification for biodiesel production. *Energy Convers. Manag.* **2009**, *50*, 2680–2684. [CrossRef]
18. Chouhan, A.S.; Sarma, A. Modern heterogeneous catalysts for biodiesel production: A comprehensive review. *Renew. Sustain. Energy Rev.* **2011**, *15*, 4378–4399. [CrossRef]
19. Correia, L.M.; Saboya, R.M.A.; Campelo, N.D.S.; Cecilia, J.A.; Rodríguez-Castellón, E.; Cavalcante, C.L.; Vieira, R.S. Characterization of calcium oxide catalysts from natural sources and their application in the transesterification of sunflower oil. *Bioresour. Technol.* **2014**, *151*, 207–213. [CrossRef]

20. Marinković, D.M.; Stanković, M.V.; Veličković, A.V.; Avramović, J.M.; Miladinović, M.R.; Stamenković, O.O.; Veljković, V.B.; Jovanović, D.M. Calcium oxide as a promising heterogeneous catalyst for biodiesel production: Current state and perspectives. *Renew. Sustain. Energy Rev.* **2016**, *56*, 1387–1408. [CrossRef]
21. Boey, P.-L.; Ganesan, S.; Maniam, G.P.; Khairuddean, M. Catalysts derived from waste sources in the production of biodiesel using waste cooking oil. *Catal. Today* **2012**, *190*, 117–121. [CrossRef]
22. Kaewdaeng, S.; Sintuya, P.; Nirunsin, R. Biodiesel production using calcium oxide from river snail shell ash as catalyst. *Energy Procedia* **2017**, *138*, 937–942. [CrossRef]
23. Bajaj, A.; Lohan, P.; Jha, P.N.; Mehrotra, R. Biodiesel production through lipase catalyzed transesterification: An overview. *J. Mol. Catal. B Enzym.* **2010**, *62*, 9–14. [CrossRef]
24. Hama, S.; Kondo, A. Enzymatic biodiesel production: An overview of potential feedstocks and process development. *Bioresour. Technol.* **2013**, *135*, 386–395. [CrossRef] [PubMed]
25. Cervero, J.M.; Alvarez, J.R.; Luque, S. Novozym 435-catalyzed synthesis of fatty acid ethyl esters from soybean oil for biodiesel production. *Biomass Bioenergy* **2014**, *61*, 131–137. [CrossRef]
26. Tongboriboon, K.; Cheirsilp, B.; H-Kittikun, A. Mixed lipases for efficient enzymatic synthesis of biodiesel from used palm oil and ethanol in a solvent-free system. *J. Mol. Catal. B Enzym.* **2010**, *67*, 52–59. [CrossRef]
27. Andreani, L.; Rocha, J.D. Use of ionic liquids in biodiesel production: A review. *Braz. J. Chem. Eng.* **2012**, *29*, 1–13. [CrossRef]
28. Feng, Y.; Qiu, T.; Yang, J.; Li, L.; Wang, X.; Wang, H. Transesterification of palm oil to biodiesel using Brønsted acidic ionic liquid as high-efficient and eco-friendly catalyst. *Chin. J. Chem. Eng.* **2017**, *25*, 1222–1229. [CrossRef]
29. Ullah, Z.; Bustam, M.A.; Man, Z. Biodiesel production from waste cooking oil by acidic ionic liquid as a catalyst. *Renew. Energy* **2015**, *77*, 521–526. [CrossRef]
30. Banković–Ilić, I.B.; Miladinović, M.R.; Stamenković, O.S.; Veljković, V.B. Application of nano CaO—Based catalysts in biodiesel synthesis. *Renew. Sustain. Energy Rev.* **2017**, *72*, 746–760. [CrossRef]
31. Bet-Moushoul, E.; Farhadi, K.; Mansourpanah, Y.; Nikbakht, A.M.; Molaei, R.; Forough, M. Application of CaO—Based/Au nanoparticles as heterogeneous nanocatalysts in biodiesel production. *Fuel* **2016**, *164*, 119–127. [CrossRef]
32. Colombo, K.; Ender, L.; Barros, A.A.C. The study of biodiesel production using CaO as a heterogeneous catalytic reaction. *Egypt. J. Pet.* **2017**, *26*, 341–349. [CrossRef]
33. Granados, M.L.; Poves, M.Z.; Alonso, D.M.; Mariscal, R.; Galisteo, F.C.; Moreno-Tost, R.; Santamaria, J.; Fierro, J. Biodiesel from sunflower oil by using activated calcium oxide. *Appl. Catal. B Environ.* **2007**, *73*, 317–326. [CrossRef]
34. Maneerung, T.; Kawi, S.; Dai, Y.; Wang, C.-H. Sustainable biodiesel production via transesterification of waste cooking oil by using CaO catalysts prepared from chicken manure. *Energy Convers. Manag.* **2016**, *123*, 487–497. [CrossRef]
35. Niju, S.; Begum, K.M.S.; Anantharaman, N. Enhancement of biodiesel synthesis over highly active CaO derived from natural white bivalve clam shell. *Arab. J. Chem.* **2016**, *9*, 633–639. [CrossRef]
36. Dizge, N.; Keskinler, B. Enzymatic production of biodiesel from canola oil using immobilized lipase. *Biomass Bioenergy* **2008**, *32*, 1274–1278. [CrossRef]
37. Charpe, T.W.; Rathod, V.K. Biodiesel production using waste frying oil. *Waste Manag.* **2011**, *31*, 85–90. [CrossRef]
38. Luo, H.; Fan, W.; Li, Y.; Nan, G. Biodiesel production using alkaline ionic liquid and adopted as lubricity additive for low-sulfur diesel fuel. *Bioresour. Technol.* **2013**, *140*, 337–341. [CrossRef]
39. Ghiaci, M.; Aghabarari, B.; Habibollahi, S.; Gil, A. Highly efficient Brønsted acidic ionic liquid-based catalysts for biodiesel synthesis from vegetable oils. *Bioresour. Technol.* **2011**, *102*, 1200–1204. [CrossRef]
40. Ren, Q.; Zuo, T.; Pan, J.; Chen, C.; Li, W. Preparation of Biodiesel from Soybean Catalyzed by Basic Ionic Liquids [Hnmm]OH. *Materials* **2014**, *7*, 8012–8023. [CrossRef]
41. Yang, J.; Feng, Y.; Zeng, T.; Guo, X.; Li, L.; Hong, R.; Qiu, T. Synthesis of biodiesel via transesterification of tung oil catalyzed by new Brønsted acidic ionic liquid. *Eng. Res. Des.* **2017**, *117*, 584–592. [CrossRef]

42. Baskar, G.; Soumiya, S. Production of biodiesel from castor oil using iron (II) doped zinc oxide nanocatalyst. *Renew. Energy* **2016**, *98*, 101–107. [CrossRef]
43. Degirmenbasi, N.; Coşkun, S.; Boz, N.; Kalyon, D.M. Biodiesel synthesis from canola oil via heterogeneous catalysis using functionalized CaO nanoparticles. *Fuel* **2015**, *153*, 620–627. [CrossRef]
44. Wen, L.; Wang, Y.; Lu, D.; Hu, S.; Han, H. Preparation of KF/CaO nanocatalyst and its application in biodiesel production from Chinese tallow seed oil. *Fuel* **2010**, *89*, 2267–2271. [CrossRef]
45. Yang, L.; Takase, M.; Zhang, M.; Zhao, T.; Wu, X. Potential non-edible oil feedstock for biodiesel production in Africa: A survey. *Renew. Sustain. Energy Rev.* **2014**, *38*, 461–477. [CrossRef]
46. Kumar, A.; Sharma, S. Potential non-edible oil resources as biodiesel feedstock: An Indian perspective. *Renew. Sustain. Energy Rev.* **2011**, *15*, 1791–1800. [CrossRef]
47. Avhad, M.; Sánchez, M.; Bouaid, A.; Martínez, M.; Aracil, J.; Marchetti, J.M. Modeling chemical kinetics of avocado oil ethanolysis catalyzed by solid glycerol-enriched calcium oxide. *Energy Convers. Manag.* **2016**, *126*, 1168–1177. [CrossRef]
48. Kumar, D.; Ali, A. Transesterification of Low-Quality Triglycerides over a Zn/CaO Heterogeneous Catalyst: Kinetics and Reusability Studies. *Energy Fuels* **2013**, *27*, 3758–3768. [CrossRef]
49. Wei, Z.; Xu, C.; Li, B. Application of waste eggshell as low-cost solid catalyst for biodiesel production. *Bioresour. Technol.* **2009**, *100*, 2883–2885. [CrossRef]
50. Kaur, N.; Ali, A. Kinetics and reusability of Zr/CaO as heterogeneous catalyst for the ethanolysis and methanolysis of Jatropha crucas oil. *Fuel Process. Technol.* **2014**, *119*, 173–184. [CrossRef]
51. Liu, S.; Nie, K.; Zhang, X.; Wang, M.; Deng, L.; Ye, X.; Wang, F.; Tan, T. Kinetic study on lipase-catalyzed biodiesel production from waste cooking oil. *J. Mol. Catal. B Enzym.* **2014**, *99*, 43–50. [CrossRef]
52. Gog, A.; Roman, M.; Tosa, M.; Paizs, C.; Irimie, F.D. Biodiesel production using enzymatic transesterification—Current state and perspectives. *Renew. Energy* **2012**, *39*, 10–16. [CrossRef]
53. Deng, L.; Xu, X.; Haraldsson, G.G.; Tan, T.; Wang, F. Enzymatic production of alkyl easters through alcoholysis a critical evaluation of lipases and alcohols. *J. Am. Oil Chem. Soc.* **2005**, *82*, 341–347. [CrossRef]
54. Andrade, T.A.; Martín, M.; Errico, M.; Christensen, K.V. Biodiesel production catalyzed by liquid and immobilized enzymes: Optimization and economic analysis. *Chem. Eng. Res. Des.* **2019**, *141*, 1–14. [CrossRef]
55. Hamd, H.T. Kinetic Processes Simulation for Production of the Biodiesel Using Enzyme as catalyst. *J. Nat. Sci. Res.* **2016**, *6*, 17–22.
56. Li, Y.; Du, W.; Dai, L.; Liu, D. Kinetic study on free lipase NS81006-catalyzed biodiesel production from soybean oil. *J. Mol. Catal. B Enzym.* **2015**, *121*, 22–27. [CrossRef]
57. Bélafi-Bakó, K.; Kovács, F.; Gubicza, L.; Hancsók, J. Enzymatic Biodiesel Production from Sunflower Oil by Candida antarctica Lipase in a Solvent-free System. *Biocatal. Biotransform.* **2009**, *20*, 437–439. [CrossRef]
58. Jachmanián, I.; Dobroyán, M.; Moltini, M.; Segura, N.; Irigaray, B.; Veira, J.P.; Vieitez, I.; Grompone, M.A. Continuous Lipase-Catalyzed Alcoholysis of Sunflower Oil: Effect of Phase-Equilibrium on Process Efficiency. *J. Am. Oil Chem. Soc.* **2009**, *87*, 45–53. [CrossRef]
59. Purushothaman, S.; Kumar, A.; Tiwari, D.P. Effect of Feeding Calcium Salts of Palm Oil Fatty Acids on Performance of Lactating Crossbred Cows Asian-Aust. *J. Anim. Sci.* **2008**, *21*, 376–385.
60. Gui, M.; Lee, K.T.; Bhatia, S. Feasibility of edible oil vs. non-edible oil vs. waste edible oil as biodiesel feedstock. *Energy* **2008**, *33*, 1646–1653.
61. Gebreegziabher, Z.; Alemu, M.; Ferede, T.; Köhlin, G. The Economics of Biodiesel Production in East Africa: The Case of Ethiopia. In *Energy-Agro-Food Nexus in East Africa: Bioenergies in East Africa between Challenges and Opportunities*; Marco Setti, M., Zizzola, D., Eds.; ACP-EU Cooperation Programme in Higher Education (EDULINK): Lungavilla, Italy, 2016; pp. 93–108.
62. Wage-Indicator-Foundation. Salary Scale in Ethiopian Public Sector. Available online: https://mywage.org/ethiopia/home/salary/public-sector-wages (accessed on 26 March 2019).
63. NUMBEO. Cost of Living in Ethiopia. Available online: https://www.numbeo.com/cost-of-living/country_result.jsp?country=Ethiopia (accessed on 26 March 2019).
64. CostToTravel. Electricity Price in Ethiopia. Available online: https://www.costtotravel.com/cost/electricity-in-ethiopia (accessed on 12 May 2019).
65. Peters, M.S.; Timmerhaus, K.D.; West, R.E.; Timmerhaus, K.; West, R. *Plant Design and Economics for Chemical Engineers*, 4th ed.; McGraw-Hill International: Singapore, 1991.

66. Engineering, C. The Chemical Engineering Plant Cost Index. Available online: http://www.chemengonline.com/pci-home (accessed on 28 March 2019).
67. Karmee, S.K.; Patria, R.D.; Lin, C.S.K. Techno-Economic Evaluation of Biodiesel Production from Waste Cooking Oil—A Case Study of Hong Kong. *Int. J. Mol. Sci.* **2015**, *16*, 4362–4371. [CrossRef]

© 2019 by the authors. Licensee MDPI, Basel, Switzerland. This article is an open access article distributed under the terms and conditions of the Creative Commons Attribution (CC BY) license (http://creativecommons.org/licenses/by/4.0/).

Article

Preliminary Study on the Use of Biodiesel Obtained from Waste Vegetable Oils for Blending with Hydrotreated Kerosene Fossil Fuel Using Calcium Oxide (CaO) from Natural Waste Materials as Heterogeneous Catalyst

S. Ozkan [1], J. F. Puna [2,3], J. F. Gomes [2,3,*], T. Cabrita [2], J. V. Palmeira [2,3] and M. T. Santos [2]

1. Kocaeli Üniversitesi Umuttepe Yerleşkesi, 41380 Kocaeli, Turkey; silaozkann@gmail.com
2. Área Departamental de Engenharia Química, Instituto Superior de Engenharia de Lisboa, Instituto Politécnico de Lisboa, R. Conselheiro Emídio Navarro 1, 1959-007 Lisboa, Portugal; jpuna@deq.isel.ipl.pt (J.F.P.); tcabrita@deq.isel.ipl.pt (T.C.); vpalmeira@deq.isel.ipl.pt (J.V.P.); tsantos@deq.isel.ipl.pt (M.T.S.)
3. CERENA—Centro de Recursos Naturais e Ambiente, Instituto Superior Técnico, Universidade de Lisboa, Av. Rovisco Pais 1, 1049-001 Lisboa, Portugal
* Correspondence: jgomes@deq.isel.ipl; Tel.: +351-96-390-2456

Received: 23 September 2019; Accepted: 7 November 2019; Published: 12 November 2019

Abstract: In this experimental work, calcium from natural seafood wastes was used as a heterogeneous catalyst separately or in a blend of "shell mix" for producing biodiesel. Several chemical reaction runs were conducted at varied reaction times ranging from 30 min to 8 h, at 60 °C, with a mass content of 5% ($W_{cat.}/W_{oil}$) and a methanol/oil molar ratio of 12. After the purification process, the biodiesel with fatty acid methyl ester (FAME) weight content measured was higher than 99%, which indicated that it was a pure biodiesel. This work also showed that the inorganic solid waste shell mixture used as the heterogeneous catalyst can be reused three times and the reused mixture still resulted in a FAME content higher than 99%. After 40 different transesterification reactions were performed using liquid (waste cooking oils) and solid (calcium seafood shells) wastes for producing biodiesel, under the specific conditions stated above, we found a successful, innovative, and promising way to produce biodiesel. In addition, blends prepared with jet fuel A1 and biodiesel were recorded with no invalid results after certain tests, at 25 °C. In this case, except for the 10% blend, the added biodiesel had no significant effect on the viscosity (fluidity) of the biojet fuel.

Keywords: biodiesel; seafood inorganic wastes; calcium oxide; transesterification; hydrotreated kerosene; heterogeneous catalysis

1. Introduction

Aviation fuel, a petroleum-based fuel used to power aircraft, has strict quality requirements in air transport [1]. Jet fuel is an aviation fuel designed specifically to power gas-turbine engines. According to a report from the U.S. Energy Information Administration in 2013, "4 gallons out of every 42-gallon barrel of crude oil are used to produce jet fuel". The worldwide aviation industry consumes approximately 1.5–1.7 billion barrels of conventional jet fuel (JET-A1) per year. Several homegrown and renewable feedstock-based fuel systems are critical in the strategy to achieve energy security and improve environmental sustainability. In 2010, the U.S. Department of Agriculture (USDA) and the Navy inaugurated a joint venture called "Farm-to-Fleet" to develop domestic, competitively priced, diesel and jet fuel replacements (USDA News Release 2013). The Farm-to-Fleet program announced in 2013 that it incorporates the acquisition of biofuel blends into regular domestic exactions for jet engine and marine diesel fuels (USDA News Release 2013). The Navy will seek to purchase JP-5 and F-76 advanced drop-in biofuels blended with 10–50% conventional fuels. There are many processual technologies to convert biomass-based materials into jet fuel [2]. Some are available on a commercial or pre-commercial scale, while others are still under research and development [3]. Global airline operations consumed approximately 1.5 billion barrels of Jet A-1 fuel producing 705 million metric tons (Mt) of CO_2 in 2013, producing just under 2% of the total of CO_2 emissions [4]. Until 2050, worldwide aviation is expecting to grow by up to 5% annually, with the following target: 12 million tons per year. CO emissions from aviation in 2012 in Europe represented 12.9% of total transport emissions. Final energy consumption in aviation in 2012 was 49.1 million tons equivalent (Mtoe) or 14% of transport energy usage. Adding biofuel to hydrotreated kerosene fossil fuel lowers the CO/CO_2 (carbon monoxide/dioxide) and NOx (nitric oxide) Green House Gas (GHG) emissions [5]. The difference between pure biofuel (B100) and fuel oil performance was 40% less for CO and 50% less for CO_2 GHG emissions. For all fuels, NOx concentration increased a little (<5%) and CO/CO_2 concentration decreased significantly. In general, previous studies suggest that, the addition of biofuel to fossil fuel reduced the static thrust and increased the fuel consumption by as much as 8% and 4%, respectively, due to the lower heat content of biofuel [6]. The presence of oxygen in biofuel molecules was expected to result in a cleaner combustion, and therefore increase the thermal efficiency, and a little increase in NOx emissions. As biodiesel is a renewable fuel, it has many advantages when compared to fossil fuels [7]. First of all, it is non-toxic, biodegradable, and does not ignite easily due to its high flash point, making it more advantageous than most fossil fuels. It is easy to transport and store it. In addition, some other advantageous reasons can be pointed out, such as, the fact that, it does not contain sulfur, it does not increase CO_2 emissions, and is produced from a renewable energy source, like the lipid biomass which can be found in oleaginous plants (rapeseed, palm, soya, sunflower, and jatropha), thus minimizing the GHG emissions [8,9]. Due to the utilization of calcium seafood wastes and vegetable oils as heterogeneous catalysts and raw materials, respectively, in biodiesel production, it promotes, twice, the sustainability of this process. Other advantages are related to the fact that biodiesel is suitable for use in diesel motor vehicles at a blend ratio less than 30%, does not require any modifications, and does not adversely affect the engine performance [10,11]. On the other hand, CO emissions are reduced by 50% and particulate matter by 30%. Sulfates, which cause acid rain, are eliminated. Aldehyde compounds are reduced by 30% and hydrocarbon emissions by 95%, when compared with the conventional diesel. In 2018, biodiesel represented 58% of the global biofuel production around the world, while bioethanol represented 39% of that production and the remaining 3% corresponded to biomethanol, biomethane, and other biofuels [12]. Figure 1 [12] shows the evolution of biodiesel and bioethanol global productions over the last 10 years. There are different methods used for biodiesel production, such as pyrolysis, dilution, transesterification, supercritical method, microwave-assisted transesterification, and ultrasound-assisted transesterification [13]. To date, different solid alkali catalysts have been developed for the production of biodiesel [14], such as zeolite, alkaline earth metal oxide [15], and hydrotalcites [16]. Alkali earth metal oxides, especially calcium oxide (CaO), are highly prominent due to their high basic strength and low solubility in methanol [17,18]. Natural Ca-rich minerals can also be used as precursors for CaO catalysts [14,19], but these materials can also be produced from seafood inorganic wastes [20,21], such as white seashells, "navalha" shells, waste obtuse horn shells [22], combusted oyster shells [23], mud crab (*Scylla serrata*) shells [24], Ca industrial wastes [25], mollusk shells, shrimp shells, and eggshells [26,27]. Basically, the activation processes of these materials only need a simple calcination process, up to 850 °C, to convert calcium carbonate into calcium oxide, the active species for these catalysts. In another study [11], it was observed how steam and carbon dioxide reacted with these materials, thus affecting the catalytic performance and behavior of active CaO. The study showed that CaO was rapidly hydrated and carbonated in air, and no CaO peak was observed after exposure to air for more than 20 days. CaO, which also shows a tendency to easily deactivate with these poisons (CO_2 and H_2O), can recover its catalytic activity if recalcination is applied. Also, several studies have been conducted to investigate the use of other natural materials as catalysts, such as snail shell [28,29] and rice husks [30], while a wide range of bio-based materials have also been reviewed recently [31]. Finally, regarding the utilization of liquid wastes to produce biodiesel, particularly waste cooking oils (WCO), by using calcium seafood wastes or the typical calcium oxide as heterogeneous catalysts, several works highlighted that higher FAME yields (>99%) are achievable with these wastes. These are very promising results for biodiesel production with high quality standards [32–36], which can improve the importance of recycling wastes, thereby decreasing the raw material costs.

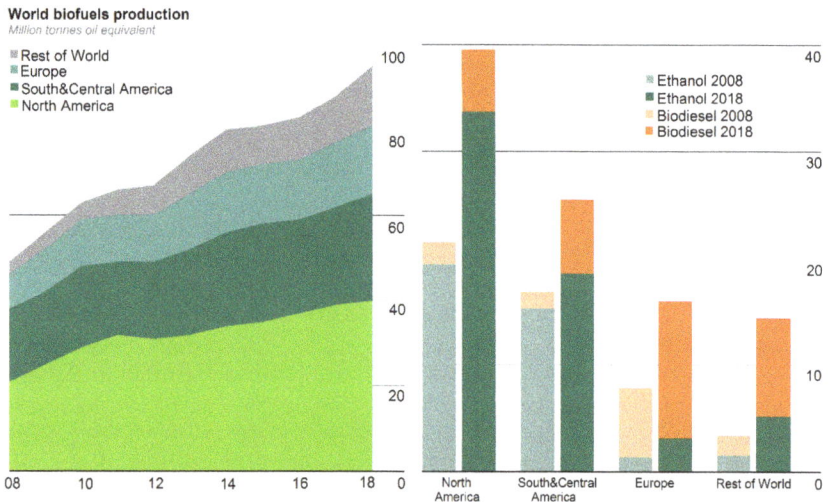

Figure 1. Global production of biodiesel and bioethanol in 2008 and 2018 [12].

2. Materials and Methods

2.1. Chemicals

The chemicals used in these experiments were phosphoric acid 85% (*w/w*) and potassium hydroxide 85% (*w/w*) from PANREAC (Barcelona, Spain); sodium hydroxide pellets from LABCHEM (Zelienople, PA, USA); citric acid; hydrated solid, pure acetone as solvent from Jose Manuel Gomes Dos Santos, Lda. (Odivelas, Portugal); ethanol >99.5% (*w/w*) and nitric acid 65% (*w/w*) from MERCK (Darmstad, Germany); methanol >99.5% (*w/w*) from CARLO ERBA (Le Vaudreuil, France); and hydrotreated kerosene (JET-A1) fuel manufactured and provided by a Portuguese refining crude oil company.

2.2. Equipment Used

Several equipment were used to achieve the various objectives of this research work as follows: Two heating baths, a LAUDA Ecoline 019 water bath with the LAUDA E100 heating system from LAUDA-Brinkmann LP (Delran, NJ, USA) and a water thermostatic bath with cooler, model F32, from JULABO Labortechnik (Seebach, Germany) as well as two mechanical stirrers from the LBX OS20 series (Labbox Labware, Barcelona, Spain) to perform the transesterification reactions; a vacuum pump from Comecta Ivymen (Comecta, Barcelona, Spain) to perform the vacuum filtration step in order to collect the catalyst samples after the transesterification process; an analytical balance, KERN EMB-V (KERN & SOHN GmbH, Balingen, Germany), to weigh the mass catalyst and prepare the solutions; an oven from ERT, Lda. (Setúbal, Portugal) to dry the catalyst samples; a furnace oven from Heraeus Instruments (Hanau, Germany) to calcinate and activate the solid catalytic samples; a centrifugation equipment, HERMLE-Z 300 (Labnet International, Inc., Edison, NJ, USA), to separate the produced soaps in the neutralization process of free fatty acids (FFA) in waste cooking oils (WCO); a heating plate with a magnetic stirrer from LBX (Labbox Labware, Barcelona, Spain) for drying biodiesel liquid samples; an FTIR-ATR (Fourier-Transform Infrared Spectroscopy-Attenuated Total Reflection) spectrometer, model Interspec 200-X, from Interspectrum OU (Tartumaa, Estonia) to plot the FTIR spectra of the solid and liquid samples; a refractometer, D'Abbé, from ATAGO (Tokyo, Japan) to quantify the FAME content of biodiesel samples produced in order to evaluate their purity; and finally, a pH/conductivity meter, AD8000, from ADWA (Szeged, Hungary) to measure the pH of washing waters in the biodiesel purification process.

2.3. Preparation of Natural CaO Catalysts

The catalyst was prepared from natural sources of $CaCO_3$. Six different calcium-rich waste components were used for the catalyst mixture. The components that made up the catalyst were eggshells, shrimp shells, crab shells, "navalha" shells, and dark-colored and white-colored seashells (clams). Shells were collected from the Lisbon beaches to prepare the Ca-based "shell mix" catalyst. The collected shells were washed and left overnight in an oven (for around 15 h), which was set to 110 °C to perform the drying process. After drying, manual grinding was completed in an agate mortar, to obtain fine powder materials. Subsequently, certain amounts of catalyst components were taken from each component and the pre-calcination process was carried out in a furnace oven at 300 °C for 3 h. After that, FTIR-ATR analysis was performed for each calcium waste material. In order to perform the calcination process, the same catalyst mixture (shell mix) was prepared through calcination at 850 °C in the same furnace oven, for 3 h. The aim was to remove the moisture and carbon dioxide contained in the catalyst mixture and convert calcium carbonate ($CaCO_3$) into calcium oxide (CaO). Thus, the catalyst was activated. At the end of the process, the catalyst contained in the ceramic capsule was taken to a desiccator for cooling.

2.4. Waste Cooking Oil Pre-Treatment

The waste cooking oils (WCO) were collected from the university canteen of ISEL (Instituto Superior de Engenharia de Lisboa), and their quality was improved through several purification steps, including drying, washing, centrifugation and neutralization. First, the WCO samples were filtered under vacuum filtration, distributed to specifically remove some immiscible solid particles, and then dried at 110 °C for 90 min to remove some humidity content. Then, the samples were placed in a heating bath at 45 °C for 15 min under mechanical stirring and at this time, approximately 0.05 g of 85% (w/w) of phosphoric acid was added to each 100 g of WCO. After this step, the acidity index (AI) was calculated and since the calculated value was found to be higher than 0.5 mg KOH/g oil, a neutralization process was performed with 8% (w/w) of an aqueous solution of sodium hydroxide (NaOH) at 45 °C for 30 min, until an AI value below 0.5 mg KOH/g oil was reached. After the neutralization process, the WCO samples were placed in a centrifugation equipment, to separate the soaps produced during the neutralization reaction, at 4000 rpm for 10 min. Then, several washing steps were performed with a citric acid aqueous solution and hot demineralized water in a decantation funnel to remove remaining contents of NaOH and some soluble impurities in the WCO samples as well as remaining contents of citric acid from the previous washings. Finally, the WCO samples were dried at 120 °C for 2 h. At this stage, the WCO samples were prepared for performing the transesterification reaction process.

2.5. Transesterification Reaction Process

The transesterification reaction procedure with the heterogenous catalyst was performed as follows: A 500 mL triple-necked flask was used as a reactor for the transesterification process. A double blade mixer apparatus was placed inside the flask and mounted on the mechanical stirrers. A cooler was used to prevent the evaporation of methanol during the reaction and it was connected to one of the reactor inlets (reflux apparatus). The set-point temperature of the water bath was set in order to achieve 60 °C inside the reactor. When the desired temperature was reached, methanol and WCO at a molar ratio of 12:1 and 5% ($W_{cat.}/W_{WCO}$) were prepared and added to the reactor. Then, a purified WCO sample, previously heated at 60 °C with the help of a heating plate, was added to the reactor and the reaction was started. Different reaction times were studied and 6 h was found to be enough as the reaction time. After changing the composition of the catalyst, according to the transesterification reaction results, 2 h was considered as the optimum reaction time. At the end of the transesterification process, the catalyst was collected on a filter paper under vacuum filtration and separated from the final liquid product, which was settled in a decantation funnel to separate the two immiscible phases: the upper one with fatty acid methyl esters (FAME—biodiesel) and the lower one with glycerin (glycerol plus unreacted methanol).

A homogeneous process was used to benchmark the heterogeneous process, and it was performed with the same operational procedure, but with the following operating conditions: 60 °C, 2 h of reaction time, a methanol/WCO molar ratio equal to 6, and 0.6% (w/w) of sodium hydroxide catalyst solubilized in methanol, related to WCO. Table 1 shows the comparison between the operating conditions used in the heterogeneous and homogeneous processes.

Table 1. Comparison between the operating values used in both heterogeneous and homogeneous catalyzed processes.

Element	Heterogeneous Process	Homogeneous Process
T (°C)	60	60
t (h)	2	2
MeOH/WCO molar ratio	12	6
Catalyst	Shell mix (CaO), c. 850 °C	NaOH with methanol
% ($W_{cat.}/W_{WCO}$)	5.0%	0.6%

2.6. Biodiesel Purification Process

After the separation of biodiesel and glycerin phases, the produced biodiesel was washed three consecutive times: the first time with demineralized water to remove non-reacted methanol and other contaminants, the second time with 1.5% (w/w) aqueous nitric acid solution to remove the remaining contents of the catalyst, and the last time with demineralized water to remove the remaining contents of nitric acid from the previous washing step. After that a centrifugation step was performed and finally, biodiesel was dried in a heating plate at 110 °C for 40 min.

3. Results and Discussion

3.1. Catalyst Characterization

Catalyst characterization was performed on natural calcium waste materials, essentially through SEM-EDS (scanning electronic microscopy with electron diffraction spectroscopy), for morphological characterization; XRD (X-ray diffraction) for structural characterization and identification of crystalline phases; and N_2 adsorption at 77 K for textural characterization to quantify solid specific area. Finally, semi-quantification of catalyst basicity was also performed through the utilization of a Hammett indicator.

Regarding morphological characterization, Figure 2 shows, from left to the right, SEM images of the shell mix new composition heterogeneous catalyst, respectively, before calcination (A), after calcination (B), and after first batch transesterification reaction step (C). All images were acquired at 8000× magnification.

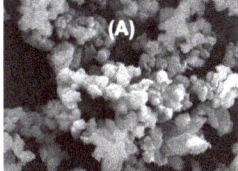 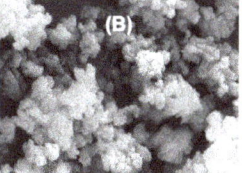 

Figure 2. SEM images (5000×, 1 μm) of shell mix solid catalyst, (**a**) before calcination, (**b**) after calcination, and (**c**) after the first transesterification step.

These images were almost similar to the CaO catalyst images reported in several works, such as [37]. Table 2 shows the atomic composition of the shell mix catalyst from situations A, B, and C, collected through the EDS technique.

Table 2. Atomic compositions (%) of shell mix ("Shellm.") collected through EDS (Energy Dispersive Spectroscopy).

Element	"Shellm." before Calcination	"Shellm." after Calcination	"Shellm." after First Reaction
C	21.6 ± 1.0%	4.39 ± 1.1%	8.1 ± 1.0%
O	59.3 ± 1.3%	59.9 ± 1.1%	63.5 ± 1.2%
Ca	19.1 ± 0.9%	35.7 ± 1.0%	28.4 ± 1.1%

Note: average values collected from three different catalyst surface measurements.

The significant decrease in carbon composition after calcination is due to the conversion of calcium carbonate to calcium oxide, which is to be expected, since this decomposition process of calcium carbonate is well known to occur between 800–900 °C. As a consequence, calcium composition increases proportionally. On the other hand, carbon composition slightly increases, due to the adsorption of oily species during the transesterification reaction, as reported in Section 3.3, covering progressively, the available active sites of the catalyst surface. Also, the noticed decrease in calcium content, after the first reaction is probably due to leaching of this element, as noticed previously in other studies [19].

Regarding catalyst crystallinity, Figure 3 shows diffractograms of the shell mix calcinated at 850 °C, before (a) and after (b) the first batch transesterification step. The differences in both diffractograms are related to the diffraction lines pointed out in diffractogram (b), ascribable to calcite (calcium carbonate), while the remaining diffraction lines are common in both diffractograms, ascribable to lime (calcium oxide). This fact seems to be related to the transformation of some of the calcium oxide particle catalyst into calcium carbonate, due to the contact of the calcium oxide catalyst with the reaction organic species. As mentioned before, calcium oxide is strongly hygroscopic and a strong CO_2 adsorber. It is also possible to notice the crystalline phases of the calcinated shell mix catalyst.

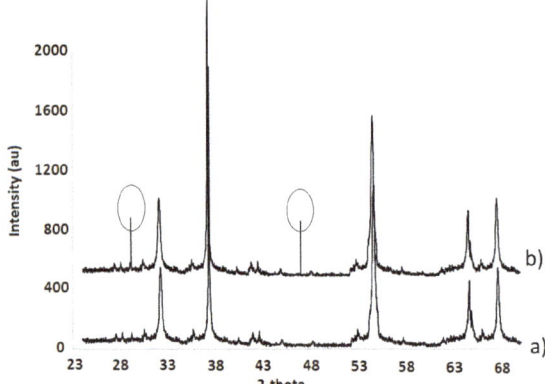

Figure 3. XRD diffractograms of the shell mix solid catalyst before (**a**) and after (**b**) the first reaction step.

An N_2 adsorption technique was applied for shell mix textural characterization at 77 K and the Brunauer–Emmett–Teller (BET) isothermal model was applied to estimate the specific area of the solid mixture. Figure 4 shows adsorption/desorption isothermal lines applied for the shell mix catalyst, after the calcination process. Through the BET model, the specific area calculated for the shell mix was equal to 2.93 $m^2 \cdot g^{-1}$, which corresponds to a macro porous solid due to its lower value of specific area. Besides, other researchers like [38] reported a value of 4.6 $m^2 \cdot g^{-1}$, which is a very close value, confirming these results. Several researchers pointed out that the calcium oxide catalyst is very active in transesterification reactions, mainly at the surface, due also to its lower value of specific area and higher value of porous volume [37,39,40].

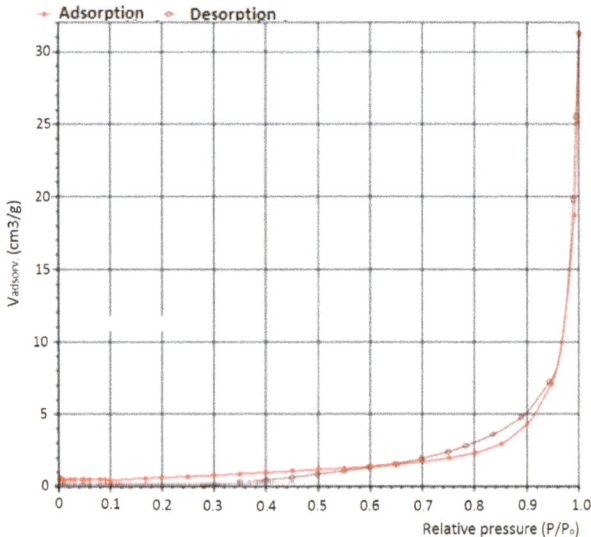

Figure 4. Adsorption and desorption isothermals of the shell mix catalyst after calcination.

The N_2 adsorption/desorption diagram of a fresh catalyst is almost the same as the one illustrated in Figure 4, with a little difference in the specific area, as the solid catalyst is characterized by a macro porous structure with a significant lower surface area, so it is not expectable to notice significative differences before and after the calcination process in these diagrams.

Figure 5 shows granulometric lines, accessed by diffraction laser-beam scattering (DLS) using methanol as a wet fluid to drag all the sample particles, representing the number of particles (%) depending on the diameter (µm) of shell mix catalyst particles before and after the transesterification reaction. It is possible to conclude that, after the reaction process, a significant number of particles with a higher diameter (e.g., 2 µm) were subjected to reduction of granulometry and converted, for instance, to 1 µm diameter. This behavior is probably due to the continuous stirring in the reactor, which causes defragmentation on the heterogeneous (solid) catalyst particles when they come into contact with the reactants' liquid phase and also due to the shovels of the mechanical stirrer, which is under a continuous rotational speed.

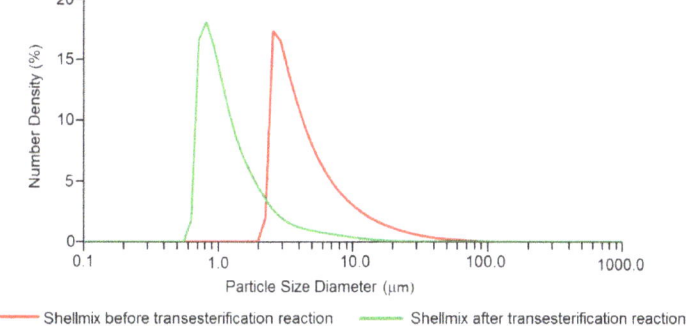

Figure 5. DLS (Dynamic Light Scattering) granulometric lines of the shell mix catalyst before and after the transesterification reaction.

Finally, a few drops of a methanolic solution of 4-chlorine-2-nitroaniline (pKa = 18.2) were used as the Hammett indicator on the shell mix solid catalyst's surface to identify its basicity. This indicator shows a yellow color in the liquid phase and also shows this same color on the catalyst surface, after deposition of the drops. After a certain time, any color change observed on the catalyst surface indicates that the catalyst is strongly alkaline with pKa > 18.2. It checks its basicity behavior, due to its alkaline active sites [41]. Figure 6 [41] shows a typical scheme of a metallic oxide, such as calcium oxide, the most important component of the shell mix catalyst, where it is possible to identify the acid (δ+) and alkaline (δ−) active sites.

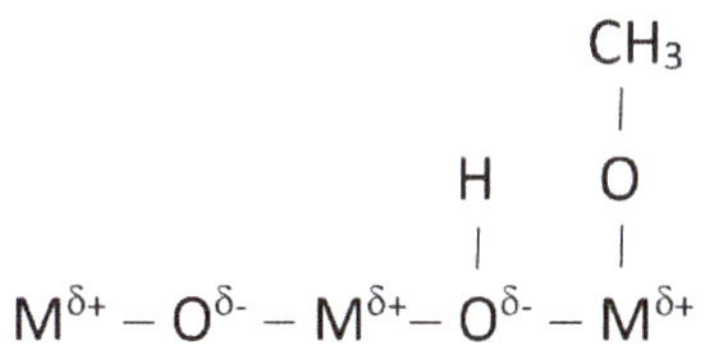

$$\begin{array}{c} \text{CH}_3 \\ | \\ \text{H} \quad \text{O} \\ | \quad | \\ M^{\delta+} - O^{\delta-} - M^{\delta+} - O^{\delta-} - M^{\delta+} \end{array}$$

Figure 6. Typical metallic oxide surface with the corresponding active sites' distribution [41].

3.2. Biodiesel Characterization

After the transesterification and separation–purification processes, the biodiesel obtained was analyzed by FTIR-ATR (Fourier-transform infrared spectroscopy) to identify the absorption peaks and compare them with the absorption peaks for homogeneously synthesized biodiesel and treated WCO (Figure 7). The homogeneous synthesis method is used as a reference for biodiesel production, in order to compare it with the heterogeneous process. The ellipse marks identified the FAME absorption peaks, which were present in the biodiesel samples, but not in the WCO samples. The biodiesel heterogeneous samples were compared with the homogeneous sample (a standard for comparison) obtained using the same reaction apparatus at 60 °C, but with a methanol: WCO molar ratio of 6:1 and 0.6% ($W_{cat.}/W_{oil}$) for 2 h, and the same purification process.

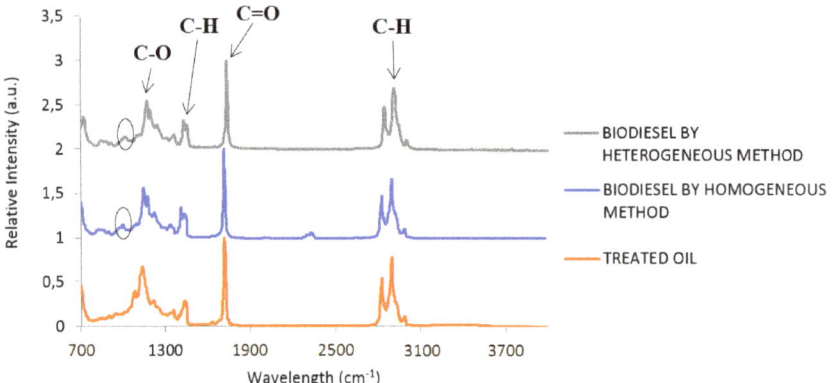

Figure 7. FTIR-ATR results for biodiesel and treated WCO (Waste Cooking Oil).

The catalyst used in the homogeneous process is, typically, sodium hydroxide solubilized in methanol. Results showed that the FTIR spectra of biodiesel homogeneous and heterogeneous samples were identical.

The composition of methyl esters in oils and in the produced biodiesel was determined by refractometry to quantify the FAME (%) weight content, which is directly related to biodiesel purity and biodiesel yield. Since soap formation was not observed in the liquid samples, it indicates that there are practically no side reactions and the FAME yield is practically equal to the transesterification conversion rate. A calibration curve was previously performed on the refractive index through the FAME (%) content, by using several standards of biodiesel in WCO (0, 20, 40, 60, 80, and 100% ($W_{biod.}/W_{WCO}$)). The obtained FAME yield results are presented in Table 3 and they showed that the calcined shell mixture of all the components initially yielded at the end of 8 h reaction time. However, considering that this is not very practical, all components were calcined again at 850 °C for 3 h in an oven and then the transesterification reaction was performed. The obtained FAME contents were then interpreted and it was concluded that the FAME percentage of white-colored clams was somewhat low. In contrast, shrimp shell, dark-colored clam, and "navalha" shell seafood wastes showed FAME yields higher than 99.5%, which is a very interesting result. However, in a previous study [37], it was reported that CaO reacted with glycerol after transesterification of soybean oil with methanol and calcium diglyceroxide was formed in the glycerin phase. Therefore, an extra purification step, for example with an ion-exchange resin, is needed to remove the soluble content in the biodiesel. Thus, in this study, egg shells may tend to react with glycerol immediately after transesterification due to its rich CaO content and form the same structure. Initially 20 g of each component (20% for each one) was used to prepare the shell mix mixture before performing the transesterification reaction for each component separately. In light of all these results, the composition, by weight, of the new shell mix mixture was changed and egg shells were removed.

The new shell mix composition, by weight, comprised of 16.7% of white-colored clams, 25.0% of dark-colored clams and crab shells, 25.0% of "navalha" shells, and 33.3% of shrimp shells. In light of the results obtained, 5% ($W_{cat.}/W_{oil}$), methanol: WCO at a molar ratio of 12:1, 60 °C working temperature, and 6 h of reaction time were considered to constitute the optimized method after performing the "repeatability test".

Table 3. FAME (Free Acid Methyl Ester) contents for biodiesel samples obtained under different conditions.

Type of Catalyst	Reaction Time (h)	Refractive Index, 25 °C	FAME Conversion (%)
Shell mix (former composition)	4	1.47735	1.10
	5	1.47731	1.20
	6	1.47729	1.30
	5 (separately calcinated)	1.46709	49.9
	2 (Methanol + catalyst, 3 h)	1.47733	1.20
	1 (Methanol + catalyst, 7 h)	1.45669	>99.5
	1 (Methanol + catalyst, 3 h) [a]	1.47725	1.80
White-colored clam	6	1.46700	50.0
Shrimp shell	6	1.45665	>99.5
"Navalha" shell	6	1.45663	>99.5
Dark-colored clam	6	1.45662	>99.5
Ca(OMe)$_2$	5	1.47730	1.20
	7	1.46716	50.8
	8	1.45661	>99.5
Ca(OH)$_2$	5	1.47727	1.40
	7	1.47726	1.40
	8	1.45661	>99.5
Glycerol + Methanol [b]	2	1.45662	>99.5
Shell mix 5%(w/w) [c]	6	1.45662	>99.5
Shell mix 3%(w/w) [c]	6	1.46690	50.4

[a] With acetone as co-solvent; [b] co-production of calcium diglyceroxide; [c] new shell mix without eggshell added. The replication measurements were only applied for those samples which showed a FAME content higher than 96.5% (w/w) (minimum value for standard quality biodiesel, according to the EN 14214 standard). In all cases, the replication shows values of FAME content higher than 99.5%.

3.3. Biodiesel Repeatability Test

After determining the most appropriate methods and conditions for biodiesel synthesis, the repeatability test was performed five times to ensure accuracy. As mentioned earlier, the same working conditions were adopted, e.g., the transesterification reaction was carried out for 6 h in a triple-necked flask using a mechanical stirrer with 5% ($W_{cat.}/W_{WCO}$) and 12:1 molar methanol: WCO ratio at 60 °C in a heated water bath (reflux apparatus), for the new shell mix composition. Repeatability test results are shown in Table 4.

Table 4. Repeatability test results for biodiesel product.

Number of Experiment (#)	FAME Yield (% $W_{biod.}/W_{WCO}$)
#1	99.9
#2	99.8
#3	99.9
#4	99.7
#5	99.8

Regarding Table 4 results, the average value achieved was 99.82% of FAME content and the standard deviation value calculated was 0.08%.

3.4. Catalytic Stability Test

Catalytic stability tests were performed to evaluate the suitability and stability of the catalyst after the first transesterification reaction (first catalytic cycle). All environment and working conditions were maintained for the transesterification reaction, e.g., tests were performed for 2 h only instead of 6 h. This change was based on kinetic test results. The catalyst, between stages, after separation of the liquid phase through vacuum filtration, was only dried overnight on a filter paper in an oven set to 90 °C. Table 5 presents the FAME content (% $W_{biod.}/W_{WCO}$) in the catalytic stability tests.

Table 5. FAME yield (%) of biodiesel products obtained as a result of catalytic stability tests.

Number of Experiment (#)	FAME Yield (% $W_{biod.}/W_{WCO}$)
#1	99.9 ± 1.0%
#2	99.9 ± 1.2%
#3	99.8 ± 0.9%
#4	28.9 ± 2.2%

As a result of these tests, it was observed that the shell mixture ("shell mix") used as the heterogeneous catalyst was suitable for use three times consecutively, which is quite an interesting result. As in another study in the literature [6], the same catalyst could be used three consecutive times with a high FAME conversion rate. The catalyst had a limited life to be reused, as seen before. In this study, it was found that the catalyst can be used three times consecutively without losing its higher activity. The reason for the significant FAME content drop from the fourth batch reaction and consequently, a decrease in the shell mix catalytic activity, is the decreasing number of catalyst active sites due to progressive adsorption of oil/methyl ester molecules on the surface of the catalyst, covering the mentioned active sites and deactivating the catalyst. For instance, the causes of deactivation for pure CaO catalyst derived from renewable resources were as follows [5]: CO_2 and H_2O existed in air and reactants, adsorbed onto the catalyst surface, poisoned it (CaO is highly hygroscopic and easily adsorbs CO_2), and converted CaO into $Ca(OH)_2$ and $CaCO_3$, respectively; the co-production of calcium diglyceroxide on the catalyst surface during reaction, which results from the interaction between CaO and glycerol; and the leaching of Ca^{2+} ions from the CaO surface, since calcium diglyceroxide is soluble in glycerin phase. Nevertheless, the catalyst can be reactivated easily, if recalcination is applied, as reported by several researchers, to desorb the oily species as well as previously adsorbed CO_2 and H_2O. However, it is important to avoid a significant formation of calcium diglyceroxide, otherwise, a catalyst weight loss with the time reaction will take place, thus leading to a significant increase in operating costs [37]. Figure 8 shows FTIR-ATR spectra of the liquid biodiesel samples obtained from the catalytic stability tests.

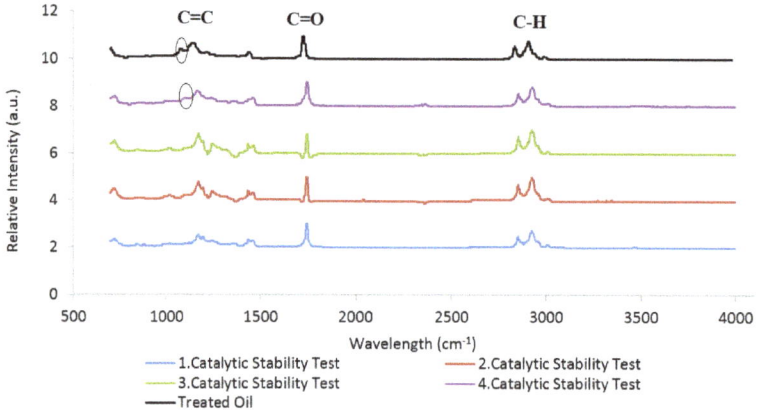

Figure 8. Comparison of FTIR-ATR results of biodiesel samples produced by catalytic stability tests and treated waste cooking oils (WCO).

3.5. Kinetic Tests

In order to evaluate time dependency, kinetic tests were performed for the new composition of the shell mix solid catalyst for different reaction times up to 6 h, at 60 °C with a methanol/WCO molar ratio of 12 and 5% of ($W_{cat.}/W_{WCO}$). The purpose was also to estimate apparent constant kinetic rates at 60 °C by using the shell mix heterogeneous catalyst. Table 6 and Figure 9 show FAME yield time dependency. After that, the apparent constant velocity (k) at 60 °C of the transesterification process was estimated for the operatory conditions mentioned in Section 2.5 (Table 1) and with the heterogeneous catalyst samples, depending on whether the reaction kinetic order law is of the 1st order or 2nd order. The best correlation coefficient (R^2) will lead to the corresponding transesterification kinetic equation. The molar ratio of methanol/WCO used corresponds to a significant excess quantity of alcohol (300%), since the stoichiometry methanol/WCO molar ratio is only equal to 3, which means that for each mol of WCO (triglyceride molecule), 3 mol of methanol will be needed in order to produce 3 mol of methyl esters (biodiesel) and 1 mol of glycerol as the co-product. This significant excess of methanol ensures that the only limiting reactant will be WCO, since in this situation, for each mol of triglyceride, 12 mol of methanol were applied, instead of the necessary (stoichiometric) 3 mol.

Table 6. Kinetic test results for different reaction times.

Reaction Time (min.)	FAME Yield (% $W_{biod.}/W_{WCO}$)
0	0.00
30	1.30
45	1.50
60	1.80
75	24.8
80	25.0
90	99.7
120	99.8
180	99.8
240	99.8
300	99.9
360	99.9

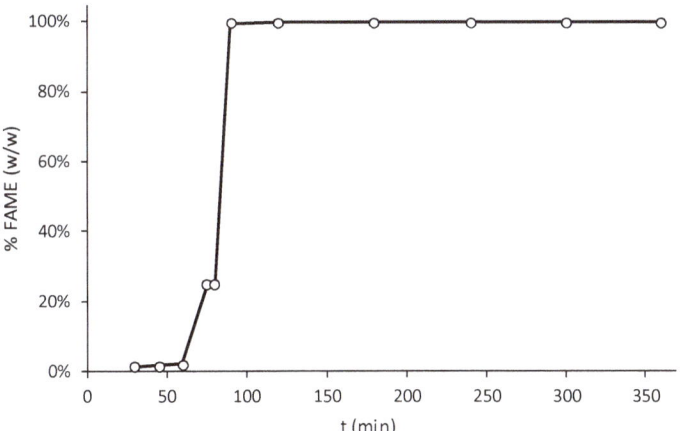

Figure 9. Kinetic test results for different reaction times.

The transesterification reaction is the result of three consecutive equilibrium reaction steps as follows: 1st step is the conversion of triglyceride (TG) molecules into diglyceride (DG) molecules; in the 2nd step, these diglycerides are converted into monoglyceride (MG) molecules; and finally, in the 3rd step, the monoglycerides are converted into glycerol (G). In each equilibrium step, 1 mol of methanol (M) is used and it will produce 1 mol of methyl ester (ME) in each step. The significant excess of methanol (300%) leads all the three equilibrium reaction steps into the direction of chemical production of reaction products, thus enhancing biodiesel production and WCO conversion, according to the Le Chatelier postulate.

1st step:
$$TG + M \leftrightarrow ME + DG.$$

2nd step:
$$DG + M \leftrightarrow ME + MG.$$

3rd step:
$$MG + M \leftrightarrow ME + G.$$

The overall reaction is given by:
$$TG + 3M \leftrightarrow 3ME + G.$$

An excess of methanol will also tend to minimize mass transfer external diffusion limitations between reactants' liquid phase and catalyst surface. Regarding mass transfer internal diffusion limitations, it was assumed that there are very few mass transfer internal diffusion limitations in this process since the catalyst is a macro porous structure and the pores have a higher diameter, thus leading to a lower surface area (<5 m$^2\cdot$g^{-1}) as mentioned before.

Those were the assumptions to define the present kinetic model, assuming pseudo-first order kinetic law, due also to the mentioned mass transfer diffusion limitations, typically occurring in heterogeneous catalysis.

From Figure 9, it is possible to conclude that there are some mass transfer limitations due to time delay in the transesterification reaction, since only after 1 h, a significant increase in FAME yield took place, when the reaction started. The rapid increase in FAME yield over time, resulted in achieving values higher than 99% at the end of only 30 min, leading to the conclusion that the reaction step is very fast and the kinetic limiting step is probably related to the mass transfer external diffusion limitations.

After performing the linearization of 1st and 2nd kinetic order equations, ($-dC_{oil}/dt = k \cdot (C_{oil})$ and $-dC_{oil}/dt = k \cdot (C_{oil})^2$, respectively, where dC_{oil} is the limiting concentration of the WCO over time (dt) and k is the apparent kinetic rate constant), the best correlation coefficient was achieved with the linearization of the 1st kinetic order equation, thus leading to the following kinetic equation:

$$-\frac{dC_{wco}}{dt} = 0.0036 \cdot C_{WCO} \qquad (1)$$

where the achieved value of k was found to be equal to 3.6×10^{-3} min^{-1}. Figure 10 shows FTIR-ATR spectra of some collected shell mix catalysts after the kinetic transesterification test. Results showed that, with time, oily species will adsorb onto the surface catalyst, since absorption peaks for the C–H and C–O bonds of esters were identified. The carbonate group, resulting from the progressive adsorption of CO_2 molecules and which reacts with calcium oxide to produce calcium carbonate, was also identified.

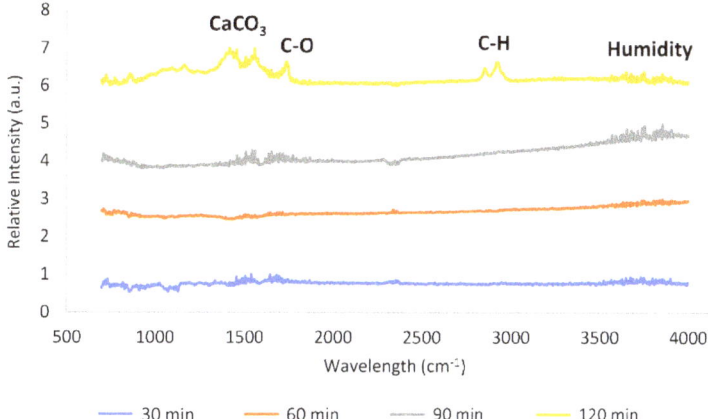

Figure 10. FTIR-ATR results of shell mix catalyst samples dried after kinetic tests.

3.6. Determination of Physical Properties of Produced Biodiesel

The physical properties of the obtained biodiesel product such as kinematic viscosity at 40 °C, density, and acidity index, besides FAME content (%), were determined and compared with the European standard EN 14214. To quantify the acidity index of the biodiesel samples, 0.1 M of KOH ethanolic solution was prepared as the titration agent through an acid-base titration. Biodiesel samples produced from the catalytic stability tests were used. The biodiesel samples were solubilized in a small volume of acetone and titrated with the KOH solution. After titration, a pink-violet color change was observed due to the presence of phenolphthalein indicator. The analysis method performed was exactly the same method used to quantify the acidity index (AI) in the WCO samples, before the neutralization process. The results are listed in Table 6. The expression for calculating the acidity index values is:

$$\text{AI (mg KOH/g oil)} = \left(\frac{56.11 \times 0.1 \times V_{KOH}}{m_{WCO}} \right), \quad (2)$$

where $V_{(KOH)}$ is the volume of KOH necessary to titrate the WCO sample, m_{WCO} is the mass of WCO weighted and used in the titration process analysis, 56.11 is the molar mass of KOH, and 0.1 is the molar concentration of the KOH solution used.

Samples 1–4 of Table 7 were biodiesel products obtained after catalytic stability (#1 = product obtained after the first catalytic stability test), while #5 sample was biodiesel produced using the homogeneous catalyst as a reference for comparison. Since the limit value of acidity index is 0.50–0.60 mg KOH/g oil, it is possible to conclude that all the obtained samples were in accordance with the biodiesel quality EU standards (EN 14214). In order to compare the biodiesel obtained from the experimental method with the commercial diesel, a density measurement was performed with the help of a pycnometer. Measurements were made at 25 °C and 40 °C only, which are close to room temperature. The results are shown in Table 8.

Table 7. Acidity index (AI) values of biodiesel products obtained from catalytic stability tests.

Biodiesel Samples	Acidity Index (AI) Control 1	Acidity Index (AI) Control 2	Average Acidity Index (AI)
#1	0.57	0.56	0.57 ± 1.3%
#2	0.55	0.53	0.54 ± 2.6%
#3	0.42	0.44	0.43 ± 1.6%
#4	0.56	0.55	0.56 ± 1.3%
#5	0.55	0.56	0.56 ± 1.3%

Table 8. Density of all samples (biodiesel, blended biodiesel-treated WCO, and blended biodiesel-hydrotreated kerosene (HK)) at 25 °C and 40 °C.

Sample	Density (kg/m³)	
Biodiesel by homogeneous method	865.0 (25 °C)	
Biodiesel by heterogeneous method (first try of catalytic stability test)	890.5 (25 °C)	842.2 (40 °C)
Biodiesel by heterogeneous method (second try of catalytic stability test)	889.9 (25 °C)	839.5 (40 °C)
Biodiesel by heterogeneous method (third try of catalytic stability test)	892.7 (25 °C)	841.3 (40 °C)
Biodiesel by heterogeneous method (four try of catalytic stability test)	919.8 (25 °C)	868.6 (40 °C)
Biodiesel according to EN 14214	**860–900 (15 °C)**	
Treated oil	925.0 (25 °C)	877.6 (40 °C)
Jet-A1	793.8 (25 °C)	
Blend of HK fuel (1% biodiesel additive)	793.6 (25 °C)	
Blend of HK fuel (2% biodiesel additive)	794.8 (25 °C)	
Blend of HK fuel (3% biodiesel additive)	795.0 (25 °C)	
Blend of HK fuel (5% biodiesel additive)	796.2 (25 °C)	
Blend of HK fuel (10% biodiesel additive)	801.9 (25 °C)	
Jet-A1 specifications	**775.0–840.0 (15 °C)**	

The same procedure was applied for jet fuel (JET-A1) and blended biodiesel–JET-A1 samples. For this test, at first the pycnometer weight was measured when empty (M0), then the pycnometer was filled with distilled water (reference fluid) and heated in a bath set at 40 °C for 5 min. After heating, the mass of water plus the pycnometer (MH$_2$O) was measured and recorded again. The same procedure was then applied to biodiesel samples (M$_{mixture}$) obtained from the catalytic stability tests and also to the blended samples of 1%, 2%, 3%, 5%, and 10% (W$_{biodiesel}$/W$_{jet\ fuel\ A1}$). Table 8 shows these results. The calculation of density (ρ) values was made through Equation (3):

$$\rho = \left(\frac{M_{mixture} - M0}{MH_2O - M0}\right) \cdot \rho H_2O \tag{3}$$

Measurements for biojet fuel blends were carried out at room temperature since the jet fuel has a flash point of 38 °C. Ambient temperature at operation was 25 °C. According to these results, all biodiesel samples from the catalytic stability tests were in accordance with the quality parameters defined for biodiesel under the European standard EN 14214, which defines density limits between 860–900 kg·m^{-3}. Regarding JET-A1, quality standards for this jet fuel define a range of density between 775–840 kg·m^{-3}, so to add biodiesel to JET-A1 fuel until achieving 10% (v/v), basically keeps the density of these mixtures within the same quality limits, which means that no relevant differences were observed. The water bath was also used to quantify kinematic viscosity at 40 °C and the capillary viscosimeter was fixed in the water bath. The spilled product from the large end of the viscosimeter was brought to the upper phase line level by means of a pump to cross to the other side. After the pump was removed, the timer was started. The timer was kept and time was recorded until the liquid level reached the lower phase line. For the same reason, measurements were made at 25 °C for the biojet-fuel-blended samples. The values of kinematic viscosity at 25 °C and 40 °C are highlighted in Table 9. The equation used for calculating kinematic viscosity values was Equation (4):

$$\frac{\mu}{\mu_{oil}} = \frac{t \cdot \rho}{t_{oil} \cdot \rho_{oil}}. \tag{4}$$

Table 9. Kinematic viscosity of all samples (biodiesel and blend of biodiesel, treated oil, and HK).

Sample	Kinematic Viscosity (mm$^2 \cdot$s^{-1}), 40 °C
Biodiesel by homogeneous method	3.50
Biodiesel by heterogeneous method (first try out of catalytic stability test)	3.72
Biodiesel by heterogeneous method (second try out of catalytic stability test)	3.52
Biodiesel by heterogeneous method (third try out of catalytic stability test)	3.60
Biodiesel by heterogeneous method (four try out of catalytic stability test)	12.6
Biodiesel according to EN 14214	**3.5–5.0**
Treated oil	32.9
Jet-A1	1.00 (*)
Blend of HK fuel (1% biodiesel additive)	1.03 (*)
Blend of HK fuel (2% biodiesel additive)	1.04 (*)
Blend of HK fuel (3% biodiesel additive)	1.07 (*)
Blend of HK fuel (5% biodiesel additive)	1.10 (*)
Blend of HK fuel (10% biodiesel additive)	1.20 (*)

(*)—Measures performed at 25 °C and these specific values represent the ratio ($\mu/\mu_{jet\text{-}A1}$) at this temperature.

Biodiesel sample viscosity results showed that all analyzed samples from the catalytic stability tests, with the exception of the fourth sample, were in accordance with the quality limits defined by EN 14214, as kinematic viscosity of FAME biodiesel must be between 3.5 and 5.0 mm$^2 \cdot$s^{-1} or cSt. The fourth transesterification batch showed a higher value of viscosity, which was related to the low value of FAME content ($\approx$28%) achieved and far from the standard quality limits.

Lower FAME content meant less biodiesel and high oil content, because the viscosity of WCO was significantly higher than the biodiesel. On the other hand, for the blended biodiesel samples in the HK fuel, the ratio between the kinematic viscosity of a blended sample (μ) and the kinematic viscosity of pure HK ($\mu_{jet\text{-}A1}$) at 25 °C slightly increased with an increase in blended FAME content (%). This correlation was linear and it was calculated with the blended samples prepared, 1%, 2%, 3%, 5%, and 10% ($W_{biod.}/W_{jet\text{-}A1}$) of the heterogeneous biodiesel produced, and HK fuel supplied by the Portuguese refining oil company. This linear correlation is presented in Figure 11. A maximum increase of 10% in kinematic viscosity was observed for the blended biodiesel up to 5% (by weight) in the HK fossil fuel, which meant that significant changes in HK viscosity and lubricity were not observed. Beyond this point, the kinematic viscosity of HK fuel, at 25 °C, started to increase significantly, thus compromising its lubricity. Further studies at significant lower temperatures must be performed to understand blended biodiesel in the lubricity behavior of hydrotreated kerosene (HK).

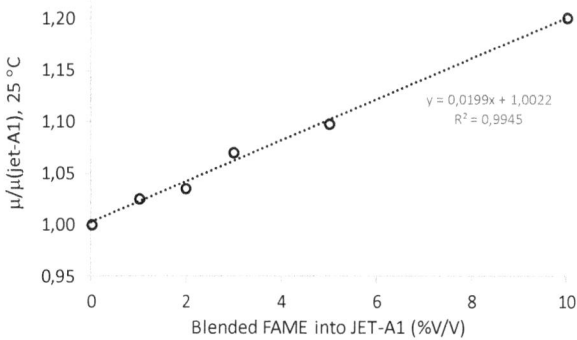

Figure 11. Linear correlation between $\mu/\mu_{(jet\text{-}A1)}$ at 25 °C and blended biodiesel in JET-A1 (HK).

4. Conclusions

At the end of this study, it is possible to conclude that waste cooking oils (WCO) can be used in order to valorize an important liquid waste, thus reducing GHG emissions, and also to use renewable primary energy sources (biomass) instead of non-renewable sources (diesel from crude oil). WCO used as raw materials have been improved and in this research work, this concept was achieved, since it was possible to produce FAME biodiesel from treated WCO and Ca-rich seafood wastes as alkaline heterogeneous catalysts. The catalytic activity and stability tests prove that these materials are suitable to be used as solid catalysts in biodiesel production, avoiding the current disadvantages of the catalytic homogeneous processes. Nevertheless, the acidity index of the WCO must be lower, otherwise, soap formation (from undesirable saponification reactions) will decrease biodiesel mass yield, since the separation process of these soaps will drag along a significant quantity of biodiesel. For that reason, WCO was pre-treated previously, with drying and neutralization processes. If not, this will affect the whole process, slow down the catalyst activity, and also decrease biodiesel purity, thereby reducing FAME yield and it will present difficulties in the separation and purification processes. It may even cause gelification of the liquid product, making it impossible to reach the desirable purity due to the undesirable production of calcium diglyceroxide, as reported before. Furthermore, by using a natural and calcium-rich heterogeneous catalyst, the biodiesel operating costs can be considerably reduced and the biodiesel can even be considered as a good alternative, since these catalyst components are easy to find and use. During the transesterification reaction, the best operating conditions achieved were 5% ($W_{cat.}/W_{oil}$), 12:1 methanol/WCO molar ratio, 60 °C, and 2 h of reaction time, thus producing a remarkable FAME conversion result higher than 99.5%, similar to current catalytic alkaline homogeneous processes with NaOH, KOH, or NaOMe. The repeatability and catalytic stability tests showed that the activity and stability of these Ca-rich waste materials made them very good catalysts in the production of FAME biodiesel by heterogeneous catalysis processes. To ensure the quality of synthesized heterogeneous biodiesel samples, some quality parameters were quantified, according to the European standard for biodiesel (EN 14214), like density, kinematic viscosity, acidity index, and FAME content. These parameters were applied for the four samples of the catalytic stability tests and with the exception of the fourth sample, the other samples showed that all values were in accordance with the standard limits, including the FAME content, which was found to be higher than 99.5% (the established standard minimum is 96.5%). The preparation of current HK fuel and FAME biodiesel blends was an important task to quantify the viscosity of these mixtures. 1%, 2%, 3%, 5%, and 10% synthesized heterogeneous FAME biodiesel samples were added to HK fuel and compared with pure JET-A1 viscosity. The data obtained from the compatibility tests were found to be the same for almost all the times recorded. Only for the blend of 10% (biodiesel), the difference was 11 s more. So, it is possible to conclude that, with the exception of the 10% blend, the added biodiesel between 1% and 5% had no significant effect on the fluidity of the HK fuel, at 25 °C. According to the obtained results, the increment of kinematic viscosity at 25 °C in the blended biojet fuel was only approximately 0.02 for each 1% (v/v) FAME biodiesel added to the HK fuel, which was a very low increase. In further studies, there is a need to perform more significant compatibility studies of blended FAME biodiesel into JET-A1 fuel, at other temperatures, especially, with very cold temperatures (−20 °C until −50 °C).

Author Contributions: Conceptualization, J.F.G. and J.F.P.; execution and discussion of the results, S.O. and T.C.; writing—review and editing, J.F.G., J.F.P., M.T.S. and J.V.P.

Funding: This research was funded by FCT (Fundação para a Ciência e Tecnologia), grant number PTDC/EMS-ENE/4865/2014.

Acknowledgments: The authors thank Maria Celeste Serra from CEEQ/ISEL for equipment availability; Beatriz Alexandre and Catarina Chaves from CLC for supplying HK fossil fuel liquid sample from the Portuguese refining crude oil company, GALP Energia. The authors also thank Isabel Nogueira from MicroLab at IST-UL for collecting SEM images and EDS data, and Ana Ribeiro from CQE/IST-UL for XRD diffractograms and for the use of the N_2 adsorption equipment. The authors are also thankful to PHARMALAB lab at ISEL and João Costa for acquiring the DLC lines.

Conflicts of Interest: The authors declare no conflicts of interest.

References

1. Wang, W.; Tao, L.; Markham, J.; Zhang, Y.; Tan, E.; Batan, L.; Warner, E.; Biddy, M. *Review of Biojet Fuel Conversion Technologies*; National Renewable Energy Laboratory: Golden, CO, USA, 2016.
2. Deane, P.; Shea, R.; Gallachoir, B.; Buchholz, D. *Biofuels for Aviation*; Rapid Response Energy Brief; Insight-E: Brussels, Belgium, 2015.
3. Korres, D.M.; Karonis, D.; Lois, E.; Linck, M.B.; Gupta, A.K. Aviation fuel JP-5 and biodiesel on a diesel engine. *Fuel* **2008**, *87*, 70–78. [CrossRef]
4. Hari, T.K.; Yaakob, Z.; Binitha, N.N. Aviation biofuel from renewable resources: Routes, opportunities and challenges. *Renew. Sustain. Energy Rev.* **2015**, *42*, 1234–1244. [CrossRef]
5. Tunca, I. Biyodizel hakkında herşey. *Biyodizel Derg.* **2013**, *8*, 41–49. (In Turkish)
6. Bartan, A. Biyodizel Üretiminde Heterojen Katalizör Geliştirilmesi. Master's Thesis, Gazi Üniversitesi, Ankara, Turkey, 2009. (In Turkish)
7. Puna, J.; Gomes, J.; Correia, M.J.N.; Dias, A.S.; Bordado, J.; Puna, J.; Correia, M.J.N.; Dias, A.P.S. Advances on the development of novel heterogeneous catalysts for transesterification of triglycerides in biodiesel. *Fuel* **2010**, *89*, 3602–3606. [CrossRef]
8. Puna, J.F.; Correia, M.J.N.; Dias, A.P.S.; Gomes, J.; Bordado, J. Biodiesel production from waste frying oils over lime catalysts. *React. Kinet. Mech. Catal.* **2013**, *109*, 405–415. [CrossRef]
9. Marchetti, J.M.; Miguel, V.; Errazu, A. Possible methods for biodiesel production. *Renew. Sustain. Energy Rev.* **2007**, *11*, 1300–1311. [CrossRef]
10. Demirbaş, A. Progress and recent trends in biodiesel fuels. *Energy Convers. Manag.* **2009**, *50*, 14–34. [CrossRef]
11. Boro, J.; Deka, D.; Thakur, A.J. A review on solid oxide derived from waste shells as catalyst for biodiesel production. *Renew. Sustain. Energy Rev.* **2012**, *16*, 904–910. [CrossRef]
12. BP Statistical Review of World Energy December 2018. Available online: https://www.bp.com/content/dam/bp/business-sites/en/global/corporate/pdfs/energy-economics/statistical-review/bp-stats-review-2019-full-report.pdf (accessed on 25 October 2019).
13. Ozcimen, D.; Yucel, S. Novel methods in biodiesel production. *Proc. Biofuel Eng. Process Technol.* **2011**, *8*, 353–384.
14. Shan, R.; Zhao, C.; Lv, P.; Yuan, H.; Yao, J. Catalytic applications of calcium rich waste materials for biodiesel: Current state and perspectives. *Energy Convers. Manag.* **2016**, *127*, 273–283. [CrossRef]
15. Puna, J.F.; Gomes, J.F.; Bordado, J.C.; Correia, M.J.N.; Dias, A.P.S. Biodiesel production over lithium modified lime catalysts: Activity and deactivation. *Appl. Catal. A Gen.* **2014**, *470*, 451–457. [CrossRef]
16. Gomes, J.F.; Puna, J.F.; Gonçalves, L.M.; Bordado, J.C. Study on the use of MgAl hydrotalcites as solid heterogeneous catalysts for biodiesel production. *Energy* **2011**, *36*, 6770–6778. [CrossRef]
17. Granados, M.L.; Poves, M.Z.; Alonso, D.M.; Mariscal, R.; Galisteo, F.C.; Moreno-Tost, R.; Santamaria, J.; Fierro, J. Biodiesel from sunflower oil by using activated calcium oxide. *Appl. Catal. B Environ.* **2007**, *73*, 317–326. [CrossRef]
18. Dias, A.P.S.; Puna, J.; Gomes, J.; Correia, M.J.N.; Bordado, J. Biodiesel production over lime. Catalytic contributions of bulk phases and surface Ca species formed during reaction. *Renew. Energy* **2016**, *99*, 622–630. [CrossRef]
19. Catarino, M.; Ramos, M.; Dias, A.P.S.; Santos, M.T.; Puna, J.F.; Gomes, J.F. Calcium Rich Food Wastes Based Catalysts for Biodiesel Production. *Waste Biomass Valorization* **2017**, *8*, 1699–1707. [CrossRef]
20. Semwal, S.; Arora, A.; Badoni, P.; Tuli, D. Biodiesel production using heterogeneous catalysts. *Bioresour. Technol.* **2011**, *102*, 2151–2161. [CrossRef] [PubMed]
21. Marwaha, A.; Rosha, P.; Mohapatra, S.K.; Mahla, S.K.; Dhir, A. Waste materials as potential catalysts for biodiesel production: Current state and future scope. *Fuel Process. Technol.* **2018**, *181*, 175–186. [CrossRef]
22. Lee, S.L.; Wong, Y.C.; Tan, Y.P.; Yew, S.Y. Transesterification of palm oil to biodiesel by using waste obtuse horn shell-derived CaO catalyst. *Energy Convers. Manag.* **2015**, *93*, 282–288. [CrossRef]
23. Nakatani, N.; Takamori, H.; Takeda, K.; Sakugawa, H. Transesterification of soybean oil using combusted oyster shell waste as a catalyst. *Bioresour. Technol.* **2009**, *100*, 1510–1513. [CrossRef] [PubMed]

24. Boey, P.L.; Maniam, G.P.; Hamid, S.A. Biodiesel production via transesterification of palm oil using waste mud crab (Scylla serrata) shell as a heterogeneous catalyst. *Bioresour. Technol.* **2009**, *100*, 6362–6368. [CrossRef] [PubMed]
25. Viriya-Empikul, N.; Krasae, P.; Nualpaeng, W.; Yoosuk, B.; Faungnawakij, K. Biodiesel production over Ca-based solid catalysts derived from industrial wastes. *Fuel* **2012**, *92*, 239–244. [CrossRef]
26. Viriya-Empikul, N.; Krasae, P.; Puttasawat, B.; Yoosuk, B.; Chollacoop, N.; Faungnawakij, K. Waste shells of mollusk and egg as biodiesel production catalysts. *Bioresour. Technol.* **2010**, *101*, 3765–3767. [CrossRef] [PubMed]
27. Piker, A.; Tabah, B.; Perkas, N.; Gedanken, A. A green and low-cost room temperature biodiesel production method from waste oil using eggshells as catalyst. *Fuel* **2016**, *182*, 34–41. [CrossRef]
28. Kaewdaeng, S.; Sintuya, P.; Nirunsin, J. Biodiesel production using CaO from river snail shell ash as catalyst. *Energy Procedia* **2017**, *138*, 937–942. [CrossRef]
29. Birla, A.; Singh, B.; Upadhyay, S.; Sharma, Y. Kinetics studies of synthesis of biodiesel from waste frying oil using a heterogeneous catalyst derived from snail shell. *Bioresour. Technol.* **2012**, *106*, 95–100. [CrossRef] [PubMed]
30. Li, M.; Zheng, Y.; Chen, Y.; Zhu, X. Biodiesel production from waste cooking oil using a heterogeneous catalyst from pyrolyzed rice husk. *Bioresour. Technol.* **2014**, *154*, 345–348. [CrossRef] [PubMed]
31. Abdullah, S.H.Y.S.; Hanapi, N.H.M.; Azid, A.; Umar, R.; Juahir, H.; Khatoon, H.; Endut, A. A review of biomass-derived heterogeneous catalyst for a sustainable biodiesel production. *Renew. Sustain. Energy Rev.* **2017**, *70*, 1040–1051. [CrossRef]
32. Kulkarni, M.G.; Dalai, A.K. Waste Cooking OilAn Economical Source for Biodiesel: A Review. *Ind. Eng. Chem. Res.* **2006**, *45*, 2901–2913. [CrossRef]
33. Melero, J.A.; Iglesias, J.; Morales, G. Heterogeneous acid catalysts for biodiesel production: Current status and future challenges. *Green Chem.* **2009**, *11*, 1285–1308. [CrossRef]
34. Shan, R.; Lu, L.; Shi, Y.; Yuan, H.; Shi, J. Catalyst from renewable sources for biodiesel production. *Energy Convers. Manag.* **2018**, *178*, 277–289. [CrossRef]
35. Dias, A.P.S.; Puna, J.; Correia, M.J.N.; Nogueira, I.; Gomes, J.; Bordado, J. Effect of the oil acidity on the methanolysis performances of lime catalyst biodiesel from waste frying oils (WFO). *Fuel Process. Technol.* **2013**, *116*, 94–100. [CrossRef]
36. Lam, M.K.; Lee, K.T.; Mohamed, A.R. Homogeneous, heterogeneous and enzymatic catalysis for transesterification of high free fatty acid oil (waste cooking oil) to biodiesel: A review. *Biotechnol. Adv.* **2010**, *28*, 500–518. [CrossRef] [PubMed]
37. Kouzu, M.; Kasuno, T.; Tajika, M.; Sugimoto, Y.; Yamanaka, S.; Hidaka, J. Calcium oxide as a solid base catalyst for transesterification of soybean oil and its application to biodiesel production. *Fuel* **2008**, *87*, 2798–2806. [CrossRef]
38. Vujicic, D.; Comic, D.; Zarubica, A.; Micic, R.; Bošković, G. Kinetics of biodiesel synthesis from sunflower oil over CaO heterogeneous catalyst. *Fuel* **2010**, *89*, 2054–2061. [CrossRef]
39. Khemthong, P.; Luadthong, C.; Nualpaeng, W.; Changsuwan, P.; Tongprem, P.; Viriya-Empikul, N.; Faungnawakij, K. Industrial eggshell wastes as the heterogeneous catalysts for microwave-assisted biodiesel production. *Catal. Today* **2012**, *190*, 112–116. [CrossRef]
40. Miladinović, M.R.; Krstić, J.B.; Tasić, M.B.; Stamenković, O.S.; Veljković, V.B. A kinetic study of quicklime-catalyzed sunflower oil methanolysis. *Chem. Eng. Res. Des.* **2014**, *92*, 1740–1752. [CrossRef]
41. Chouhan, A.S.; Sarma, A. Modern heterogeneous catalysts for biodiesel production: A comprehensive review. *Renew. Sustain. Energy Rev.* **2011**, *15*, 4378–4399. [CrossRef]

 © 2019 by the authors. Licensee MDPI, Basel, Switzerland. This article is an open access article distributed under the terms and conditions of the Creative Commons Attribution (CC BY) license (http://creativecommons.org/licenses/by/4.0/).

Review

Biodiesel Production Processes and Sustainable Raw Materials

Marta Ramos [1,2], Ana Paula Soares Dias [2,*], Jaime Filipe Puna [1,2], João Gomes [1,2] and João Carlos Bordado [2]

[1] ADEQ, Instituto Superior de Engenharia de Lisboa, Instituto Politécnico de Lisboa, R. Conselheiro Emídio Navarro, 1, 1959-007 Lisboa, Portugal; martasaramos@tecnico.ulisboa.pt (M.R.); jpuna@deq.isel.ipl.pt (J.F.P.); jgomes@deq.isel.ipl.pt (J.G.)
[2] LAETA, IDMEC, CERENA, Instituto Superior Técnico, Universidade de Lisboa, Av. Rovisco Pais, 1, 1049-001 Lisboa, Portugal; jcbordado@tecnico.ulisboa.pt
* Correspondence: apsoares@tecnico.ulisboa.pt

Received: 16 October 2019; Accepted: 12 November 2019; Published: 20 November 2019

Abstract: Energy security and environmental concerns, related to the increasing carbon emissions, have prompted in the last years the search for renewable and sustainable fuels. Biodiesel, a mixture of fatty acids alkyl esters shows properties, which make it a feasible substitute for fossil diesel. Biodiesel can be produced using different processes and different raw materials. The most common, first generation, biodiesel is produced by methanolysis of vegetable oils using basic or acid homogeneous catalysts. The use of vegetable oils for biodiesel production raises serious questions about biodiesel sustainability. Used cooking oils and animal fats can replace the vegetable oils in biodiesel production thus allowing to produce a more sustainable biofuel. Moreover, methanol can be replaced by ethanol being totally renewable since it can be produced by biomass fermentation. The substitution of homogeneous catalyzed processes, nowadays used in the biodiesel industry, by heterogeneous ones can contribute to improve the biodiesel sustainability with simultaneous cost reduction. From the existing literature on biodiesel production, it stands out that several strategies can be adopted to improve the sustainability of biodiesel. A literature review is presented to underline the strategies allowing to improve the biodiesel sustainability.

Keywords: biodiesel; sustainability; vegetable oils; animal fats; methanolysis; ethanolysis

---

1. World Energy

Worldwide energy demand has been growing in the last decades (Figure 1a). According to the U.S. Energy Information Administration (EIA) report, this trend will carry on, with an estimated growth in energy consumption of 28% between 2015 and 2040 [1]. Only in 2018, the world primary energy consumption grew 2.9% [2].

World use of petroleum and other fuels has been growing as well, being the largest growth in the transport and industrial sector. In the transportation sector, fossil fuels continue to supply most of the energy consumed despite the shortage of their reserves [1].

In Africa, Europe and Americas the oil remains the dominant fuel, while natural gas dominates in the Commonwealth of Independent States (CIS) and the Middle East. In the Asia Pacific region, coal is the dominant fuel (Figure 1b).

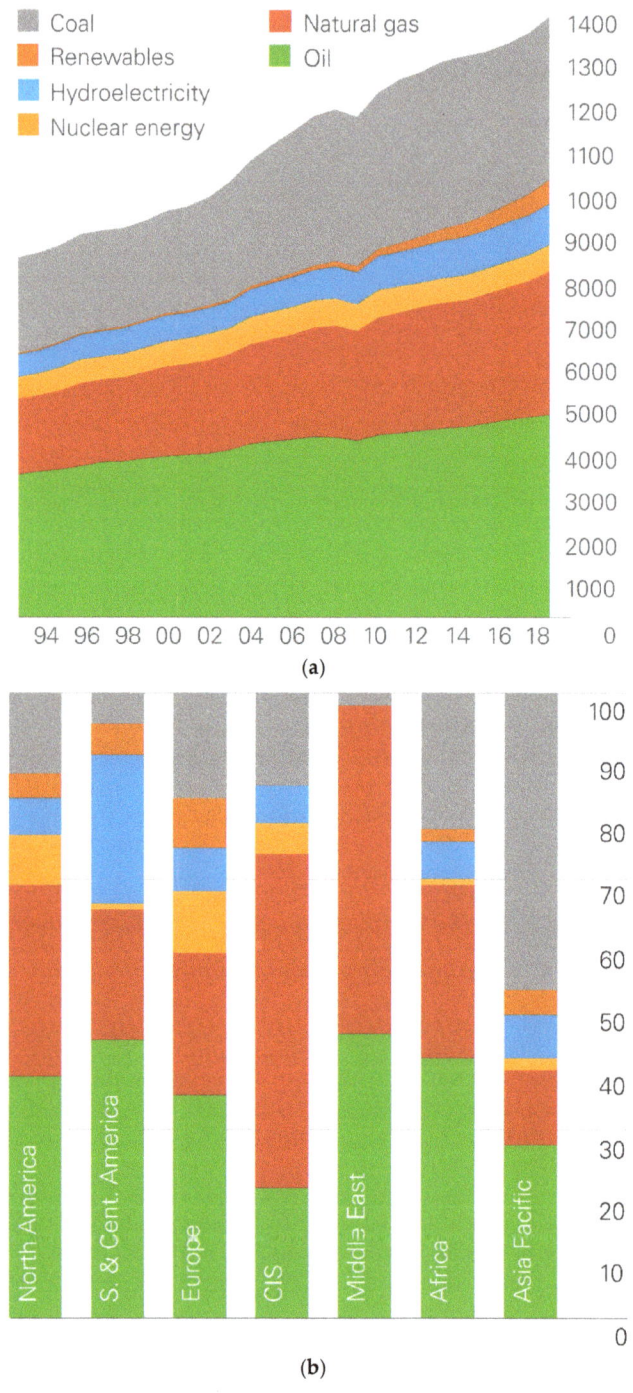

Figure 1. World Consumption by fuel from 1994 to 2018. (**a**) (million tones oil equivalent) and Regional Consumption by fuel 2018; (**b**) (% of different fuels, color legend in Figure 1a) [2].

Replacing fossil fuel with more sustainable energies, maximizing the use of renewable ones, is increasingly important, not only to reduce the emissions of greenhouse gases (GHG) but also to improve energy supply security [3]. These concerns have led to changes in global environmental policy.

In 2007 the European Union launched a climate and energy policy to fight climate change and increase energy security but reinforcing simultaneously its competitiveness. The 2020 package was enacted in legislation in 2009 (Renewable Energy Directive) and sets targets for the year 2020 [4]:

- ✓ 20% cut in greenhouse gas emissions (from 1990 levels)
- ✓ 20% of EU energy from renewable sources in the energetic mix
- ✓ 20% improvement in energy efficiency

The EU also sets binding national targets of minimum energetic incorporation of 10% for the share of energy from renewable sources consumed by all modes of transport in 2020 [5]. One way to achieve the proposed targets is the increase in the use of biofuels as an alternative energy source. Figure 2 shows the share of renewable energy in transport in 2014, 2015 and 2016 for EU countries.

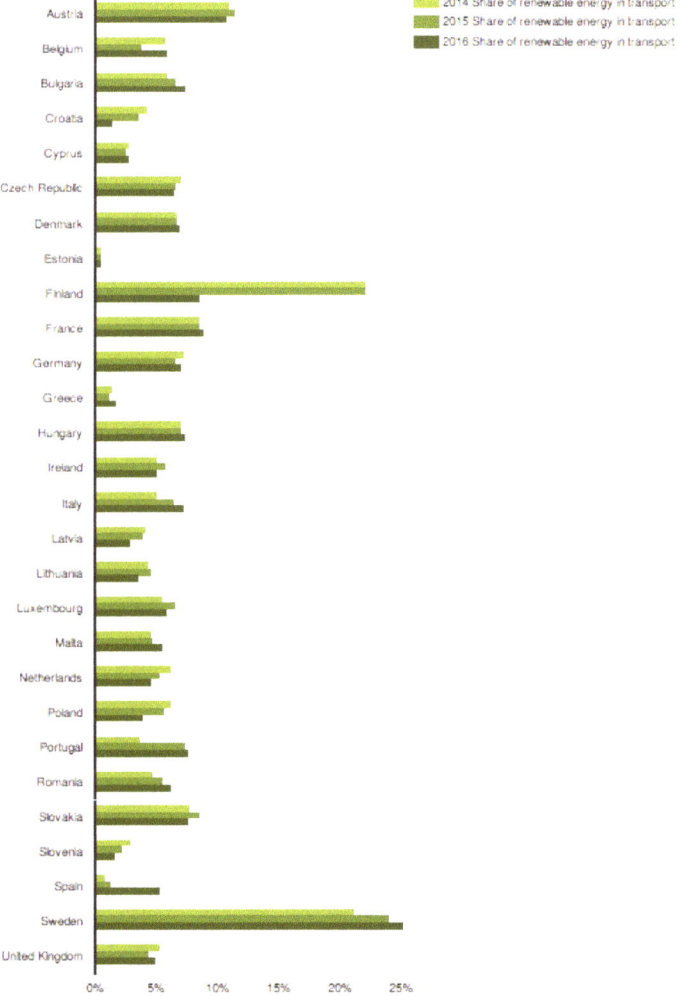

Figure 2. EU share of renewable energy in transport in 2014, 2015 and 2016 [6].

The Renewable Energy Directive was reviewed in 2015 (Directive (EU) 2015/1513) limiting to 7% biofuel production from agri-food-cultures such as cereal and other starch-rich crops, sugars and oil crops used in transport sector [7].

Biofuel production from wastes and residues was also encouraged due to double contribution by double counting for the purpose target. In addition to the current list of raw materials that can be used to produce double counted biofuels (Directive (EU) 2015/1513) it is possible to use raw materials not included in the list but considered as wastes by the national authorities before the adoption of the amendment [7]. Not all countries apply double counting and the definition of waste differs between them.

For example, Portugal is one of the countries that apply double counting, and biofuels produced from animal fats categories I & II and waste cooking oils, among others, are counted twice. The EU targets for the year 2030 (from 2021 to 2030) had been already established [8]:

- ✓ At least 40% cuts in greenhouse gas emissions (from 1990 levels)
- ✓ At least 32% share for renewable energy (upwards revision by 2023)
- ✓ At least 32.5% improvement in energy efficiency

The EU has also set a new binding national target of minimum energetic incorporation of 14% for the share of energy from renewable sources consumed in transport until 2030. The 2050 long-term strategy, instead of set targets, creates a vision and defines directions that the EU must take to achieve climate neutrality as well as the Paris Agreement, which established keeping the temperature increase well below 2 °C compared to pre-industrial levels and the pursuit of efforts to keep it to 1.5 °C, by 2050. Several strategic areas such as energy efficiency; deployment of renewables; clean, safe and connected mobility; competitive industry and circular economy; infrastructure and interconnections; bio-economy and natural carbon sinks; carbon capture and storage to address remaining emissions would have to be worked together to achieve the climate neutrality. In the transport sector an increase in biofuels production due to all alternative fuel options is predicted, which will be required achieving deep emission reductions [9].

2. Biofuels

Biofuels are fuels made from biomass, a renewable alternative to fossil fuels. Many of them can be used in the transport sector, like [10]:

- bioethanol: ethanol produced from biomass and/or the biodegradable fraction of waste;
- biodiesel: a methyl-ester produced from vegetable or animal oil, of diesel quality;
- biogas: a fuel gas produced from biomass and/or from the biodegradable fraction of waste, that can be purified to natural gas quality, to be used as biofuel, or wood gas;
- biomethanol: methanol produced from biomass;
- biodimethylether: dimethylether produced from biomass,
- bio-ETBE (ethyl *tert*-butyl ether): ETBE produced based on bioethanol. The percentage by volume of bio-ETBE that is calculated as a biofuel is 47%;
- bio-MTBE (methyl *tert*-butyl ether): a fuel produced based on biomethanol. The percentage by volume of bio-MTBE that is calculated as a biofuel is 36%;
- synthetic biofuels: synthetic hydrocarbons or mixtures of synthetic hydrocarbons, which have been produced from biomass;
- biohydrogen: hydrogen produced from biomass, and/or from the biodegradable fraction of waste;
- pure vegetable oil: oil produced from oil plants through pressing, extraction or comparable procedures, crude or refined but chemically unmodified, when compatible with the type of engines involved and the corresponding emission requirements.

In this sector, the most widely used biofuels around the world are bioethanol, as a substitute for gasoline, and biodiesel, as a substitute for diesel. Other biofuels are also used, although with more limited market access.

2.1. Biodiesel

Biodiesel, a mixture of alkyl esters produced of fatty acids is highlighted out as a feasible renewable and low carbon substitute of fossil diesel for the transportation sector [5]. Biodiesel can be used pure or blended with petroleum diesel due to its complete miscibility. Biodiesel blends are referred to as Bxx, where the xx indicates the amount of blend. Thus, B100 corresponds to pure biodiesel, and a B80 blend is 80% biodiesel and 20% petroleum diesel by volume.

Worldwide Europe is the main producer of biodiesel as a result of the environmental policy (Figure 3). Diverse feedstocks can be employed in biodiesel production. Nowadays biodiesel worldwide production is still dominated by vegetable oils: soybean, rapeseed, and palm oil. In the USA the main raw material used is soybean oi, with a 52% share of total biodiesel feedstocks, followed by canola oil and corn oil with 13% each [11]. In Europe, rapeseed oil was the major feedstock used, with 45% of the total production in 2017, followed by used cooking oil (UCO) with 21% and palm oil with 18% [12]. For example, in Portugal, rapeseed is the main vegetable oil used, followed by soybean oil. Table 1 presents Portugal's biodiesel production in the last years.

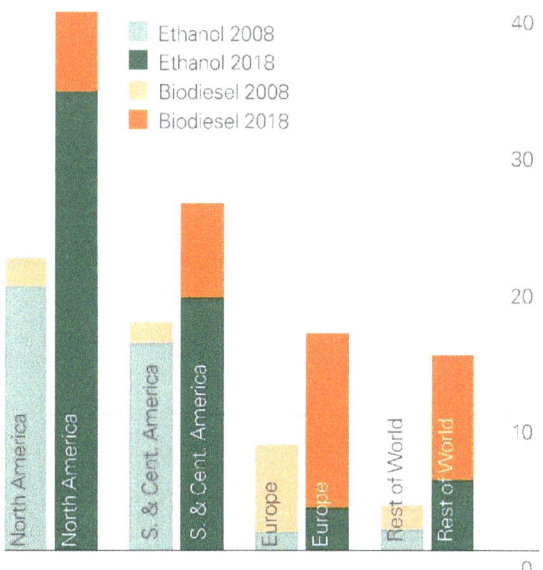

Figure 3. World ethanol and biodiesel production from 2008 to 2018 (vertical axis in million tonnes oil equivalent) [2].

Table 1. Biodiesel production from different feedstocks in Portugal (adapted from [13]).

Feedstock	Biodiesel Production (ton); Year								
	2010	2011	2012	2013	2014	2015	2016	2017	2018
Fresh oils	316,507	365,622	304,190	299,404	324,200	287,329	205,594	175,954	151,078
WCO+ animal fat	4810	4639	4869	11,044	16,906	75,737	131,226	179,875	176,023
Total	321,317	370,261	309,059	310,448	341,106	363,066	336,820	355,828	327,101

2.2. Advantages and Disadvantages

The major benefits of using biodiesel as a replacement for diesel fuel are [14–17]:

- Biodegradability;
- Non-flammable and low toxicity;
- Safer to handle;
- Higher combustion efficiency, portability, availability, and renewability;
- Higher cetane number and flash point;
- Lower emissions such as CO_2, CO, SO_2, particulate matter (PM) and hydrocarbons (HC) compared to diesel;
- May be blended with diesel fuel at any proportion;
- No required engine modification up to B20;
- Excellent properties as a lubricant.

There are also some disadvantages of using biodiesel that must be taken into consideration:

- Lower calorific value;
- Higher pour and cloud point fuel;
- Higher nitrous oxide (NOx) emissions (in some cases);
- Higher viscosity and less oxidative stability;
- Biodiesel is corrosive to copper and brass;
- May degrade plastic and natural rubber gaskets and hoses when used in pure form;
- Biodiesel causes excessive engine wear.

The main restriction for biodiesel commercialization is its higher cost in comparison to petroleum fuel. Raw materials price represents 70–95% of the total production cost [18].

2.3. Transesterification

Biodiesel is produced by transesterification of triglycerides with short-chain alcohols in the presence of a catalyst (Scheme 1). Due to the reversibility of the reaction, it is necessary to use an excess of alcohol to drive the reaction equilibrium [19]. However, the transesterification reaction can be done without a catalyst through supercritical process reactions [20]. This process consists of three consecutive reversible reactions where triglycerides are converted into diglycerides, diglycerides are converted into monoglycerides and finally, monoglycerides are converted into glycerol. In addition, for each glyceride that reacts the formation of an ester (biodiesel) molecule occurs [21].

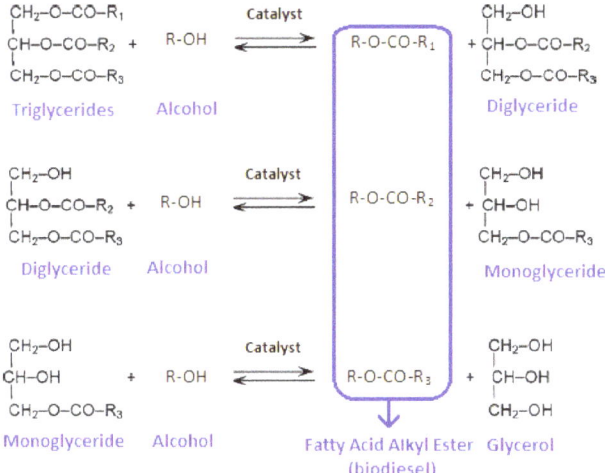

Scheme 1. Transesterification reaction (adapted from [21]).

Various parameters affect the transesterification reaction. In order to produce biodiesel that meets the standard quality parameters, production processes must be optimized [3]. The most relevant, processual and quality, parameters are [3,22,23]:

(1) Free fatty acids, moisture and water content.
(2) Type of alcohol and molar ratio employed.
(3) Type and concentration of catalysts.
(4) Reaction temperature and time.
(5) Rate and mode of stirring.
(6) Purification process of the final product.
(7) Mixing intensity.
(8) Effect of using organic co-solvents.
(9) Specific gravity

Glycerol, also known as glycerin (commercial term, purity > 95%), is a by-product of transesterification reactions. About 1 kg is produced for each 10 kg of biodiesel [24]. Glycerol is a nontoxic, edible and biodegradable compound used as a raw material in different industries, such as pharmaceuticals, cosmetics, tobacco, textiles or foods [24,25]. Due to its commercial value, in transesterification reaction beyond the biodiesel quality it is also important to obtain glycerol with high purity.

2.4. Alcohol Used

In biodiesel production, different alcohols can be used, such as methanol, ethanol, propanol or butanol [26]. The most commonly used are methanol and ethanol, and the reaction product produced when methanol is used is called a fatty acid methyl ester mixture (FAME) whereas if the alcohol is ethanol, the product obtained is a fatty acid ethyl ester mixture (FAEE) [26].

The mild reaction conditions needed, the fast reaction time and the easy phase separation combined with its low-cost and industrial availability make the methanol the most used alcohol in biodiesel production [26,27]. However, the use of this alcohol presents some drawbacks. Methanol is more toxic, volatile and has a lower oil dissolving capacity than ethanol. Although methanol can be obtained from biomass gasification, this alcohol is majorly produced from a fossil sources, about 90% from natural gas. Thus, the biofuel produced by methanolysis is not considered fully renewable biodiesel [28].

Besides, ethanol is made from agricultural products such as potatoes, grain, and corn, allowing this way the production of a renewable fuel [29]. Due to the extra carbon atom, the FAEE produced has a cloud and pour point lower than FAME, which allows the engine to start low temperatures [30]. The combustion heat and the cetane number are higher and the storage properties of FAEE fuel are also improved [27]. The main drawbacks of ethanolysis in biodiesel production are its lower reactivity, compared with methanol, as well as the more difficult separation of FAEE from the coproduced glycerin due to their higher miscibility [31].

Many studies have been carried out to compare the effect of methanol and ethanol on biodiesel production from different feedstocks [32–35]. All achieved results reported that the yield obtained by ethanolysis is lower and more time is needed to complete the reaction than for methanolysis. The separation of FAEE from glycerin is also more difficult. Nevertheless, it allows achieving a completely renewable biodiesel. Although several alcohols can be used to produce biodiesel, so far European Union legislation only covers FAME.

2.5. Feedstocks

As mentioned before several feedstocks can be employed in biodiesel production such as vegetable oils (edible and non-edible), waste cooking oils, animal fats and algae oils [36]. The chemical structure is similar in vegetable oils and animal fats, mainly composed by triglycerides with a smaller fraction of diglycerides and monoglycerides [18]. Triglycerides (Figure 4) are formed by one molecule of glycerol combined with three molecules of saturated or unsaturated fatty acid.

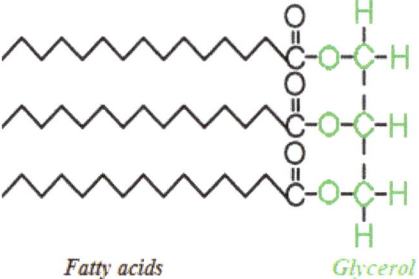

Figure 4. Triglyceride [37].

Both feedstocks are water-insoluble, hydrophobic and soluble in nonpolar organics solvents [18]. The main differences between them are the varied distributions of fatty acids and the high free fatty acids (FFA) content in the fats. The fatty acid profiles of some vegetable oils and animal fats are referenced in Table 2.

Animal fats and greases, at room temperature, tend to be solid due to their high content of saturated fatty acid (carbon-carbon single bond), oils are generally liquids. Refined oils have less FFA, lower acid value, than animal fats, waste grease and waste oils [18].

Table 2. Fatty acid profiles of different feedstocks (minima and maxima weight percentages).

Fatty Acid	Rapeseed Oil [18,38,39]	Soybean Oil [18,38,40,41]	WFO [39,41–43]	Beef Tallow [38,40,44–46]	Lard [18,38, 40,44,47]	Poultry Fat Chicken Fat [18,38,40,44,48]	Poultry Fat Duck Tallow [44]	Catfish Fat [19]	Fish Oil Salmon Oil [50]	Fish Oil Anchovy Oil [51]
Lauric (C12:0)	-	-	nm–0.4	-	-	nm–1.0	-	11.5	0.1	-
Myristic (C14:0)	-	nm–1.0	nm–1.1	2.6–3.5	1.3–1.7	0.5–1.0	-	11.7	5.8	6.7
Myristoleic (C14:1)	-	-	-	0.5–1.3	-	0.1–0.2	-	2.0	-	-
Pentadecanoic (C15:0)	-	-	-	0.5–1.0	-	-	-	1.9	-	-
Palmitic (C16:0)	3.5–4.5	10.5–11.0	8.4–25.8	23.8–27.0	23.2–25.5	20.9–24.7	17.0	28.1	16.9	20.2
Palmitoleic (C16:1)	nm–0.5	-	0.2–4.6	0.5–4.7	2.2–2.7	5.0–7.7	-	-	5.4	6.6
Margaric (C17:0)	-	-	-	1.1–2.5	nm–0.4	-	-	-	-	0.2
Heptadecenoic (C17:1)	-	-	-	0.5–1.7	nm–0.4	-	-	-	-	-
Stearic (C18:0)	0.9–1.5	3.3–4.8	3.7–4.8	12.7–34.7	10.4–17.0	4.5–5.8	4.0	-	4.3	4.2
Oleic (C18:1 cis)	-	22.0–25.4	28.5–52.9	29.9–47.2	40.0–42.8	38.2–48.5	59.4	26.8	19.2	19.7
Linoleic (C18:2)	18.7–22.3	52.3–54.5	13.5–50.5	0.8–2.7	10.7–21.0	17.3–23.8	19.6	6.7	16.1	2.6
Linolenic (C18:3)	7.7–9.0	5.3–7.5	0.6–3.5	nm–0.8	nm–64.7	nm–2.5	-	-	2.8	1.6
Arachidic (C20:0)	0.4–0.5	0.4–0.5	0.1–0.4	nm–0.3	nm–0.2	-	-	-	-	-
Gadoleic acid (C20:1)	1.0–2.0	nm–0.3	0.1–0.8	nm–0.5	0.9–1.0	0.5–1.0	-	2.7	-	-
Eicosadienoic (C20:2)	-	-	-	-	0.5–0.7	-	-	0.8	-	0.2
Eicosatrienoic (C20:3)	-	-	-	-	nm–0.2	-	-	0.5	-	-
Eicosapentaenoic (C20:5)	nm–0.1	-	0.2	-	-	-	-	-	15.6	10.4
Behenic (C22:0)	nm–0.5	-	nm–0.8	-	-	-	-	-	-	-
Erucic (C22:1)	nm–0.1	-	-	-	-	-	-	-	-	-
Docosapentaenoic (C22:5)	-	-	-	-	-	-	-	-	2.5	0.8
Docosahexanoic (C22:6)	-	-	-	-	-	-	-	-	11.4	21.6
Lignoceric (C24:0)	-	nm–0.1	0.04–0.3	-	-	-	-	-	-	-

nm—not measured.

Based on their feedstock, biodiesel can be classified into three categories: first, second and third generation (Table 3):

Table 3. Different generations of biodiesel and their feedstocks (adapted from [3]).

1st Generation Edible Oils	2nd Generation		3rd Generation Microalgal Oils
	Non-Edible Oils	Animal Fats	
Soybeans (*Glycine max*)	Jatropha (*Jatropha curcas* L.)	Pork lard	Bacteria
Rapeseed (*Brassica napus* L.)	Mahua (*Madhuca longifolia*)	Beef tallow	Microalgae (*Chlorella prothecoides*)
Safflower (*Carthamus tinctorius* L.)	Coffee grounds	Poultry fat	Microalgae (*Chlorella vulgaris*)
Rice bran oil (*Oryza sativa* L.)	Camelina (*Camelina sativa*)	Fish oil	Microalgae (*Botryococcus braunii*)
Barley (*Hordeum vulgare* L.)	Cottonseed (*Gossypium hirsutum*)	Chicken fat	Microalgae (*Chlorella sorokiana*)
Sorghum (*Sorghum bicolor*) Wheat (*Triticum aestivum*) Corn (*Zea mays*) Coconut (*Cocos nucifera*) Canola (*Brassica napus*) Peanut (*Arachis hypogaea*)	Tall fescue (*Festuca arundinacea*) Neem (*Azadirachta indica*) Jojoba (*Simmondsia chinensis*) Passion seed Moringa (*Moringa oleifera*)		
Palm (*Arecaceae*)	Tobacco seed (*Nicotiana tabacum*)		
Sunflower (*Helianthus annuus*)	Rubber tree seed (*Hevea brasiliensis*)		
Palm kernel (*Elaeis guineensis*)	Nag champa (*Plumeria*)		

First generation refers to biodiesel derived from edible vegetable oils. The most commonly used are rapeseed, palm, soybean, coconut, peanut, and sunflower [52]. The vegetable oils are widely available and relatively easier to convert into biodiesel. However, the use of edible vegetable oils in the production of biofuel raises several ethical issues. Edible vegetable oils come from food crops. The use of arable land, water, and fertilizer in "growing fuel" instead of food not only affects the food price but also sustainability issues [52].

Furthermore, even if the total amount of edible oils available was used in the production of biodiesel, it was not enough to meet today's diesel requirements. These concerns ally with the double counting of biofuels produced from wastes, which have led to an increasing search for more sustainable feedstocks.

Second generation biofuels are biodiesels derived from non-edible crops or feedstocks that have already fulfilled their food purpose such as waste oily streams from the oil refinery, waste cooking oils (WCOs), greases and waste animal fats (WAFs) [53]. The non-edible crops can be grown on lands that cannot be used for arable crops that have a lower necessity of water or fertilizer to grow, making their plantation more economic [53]. The WCO refers to vegetable oils or animal fats that had been heated and used for cooking different types of food. During this process, various chemical reactions occur such as hydrolysis, polymerization, and oxidation modifying the physical and chemical properties of oil/fat [54]. Recycled fats, based on their FFA level, can be divided as yellow or brown grease. The yellow greases have a FFA level of less than 15% while, brown has more than 15% [54]. The second generation also includes WAFs or rendered animal fats, this topic will be explored in the next chapter. The use of these less expensive feedstocks (Table 4) reduces the production costs and reuse wastes, without competing with the food market [47]. The prices of feedstocks are unstable.

Third generation are the biodiesels derived from algal biomass.

Independently of the feedstock category used, the physical and chemical properties of the biodiesel are the same [41].

Table 4. The prices of the feedstocks [55,56].

Type	Price
soybean oil	728 USD per ton [1]
rapeseed oil	827 USD per ton [1]
palm oil	535 USD per ton [1]
WCO	610 USD per ton [2]
tallow (category 1)	400 € per ton [2]

[1] December 2018 price; [2] October 2018 price.

Animal Fats

Biodiesel production can be also done with animal fats as raw materials such as tallow, lard, poultry fat and fish oils (Figure 5) [18]. Animal fats are wastes or by-products that came from animal meat processing industry and carcasses of livestock, with relatively low prices.

Figure 5. Tallow, lard, fish oil, and poultry fats.

In the European Union the regulation (EC) No 1069/2009 and No 142/2011 lays down health rules as regards animal by-products and derived products not intended for human consumption. These materials can be categorized into three specific categories considering the perceived level of risk to public and animal health [57]:

Category 1 (high risk):

✓ Specified Risk Material (SRM) linked with the transmission of TSEs (Transmissible Spongiform Encephalopathies), this includes the spinal cord and brain.
✓ Fallen stock with SRM
✓ Catering waste
✓ Anything handled with Category 1

Category 2:

✓ Material not fit for human consumption and posing a risk to animals and humans
✓ Fallen stock without SRM

Category 3 (lowest risk):

✓ Fit for human consumption at the point of slaughter

Fats are recovered from waste fat tissues by the rendering process. This process depends on the risk category and to prevent contamination between different categories of waste and different species, all processing is done on separated lines [57].

Many types of rendering are used in the industry. All of them involve the application of heat, the extraction of moisture, and the separation of fat [58]. The fat can mainly be recovered from wet or dry rendering. In wet rendering (Figure 6), the fat is recovered by heating in the presence of water. Boiling in water and/or steam at a high temperature can be employed [58,59]. The color of the fat produced by this process is clearer. The free fatty acid content increases due to the long contact with water [58].

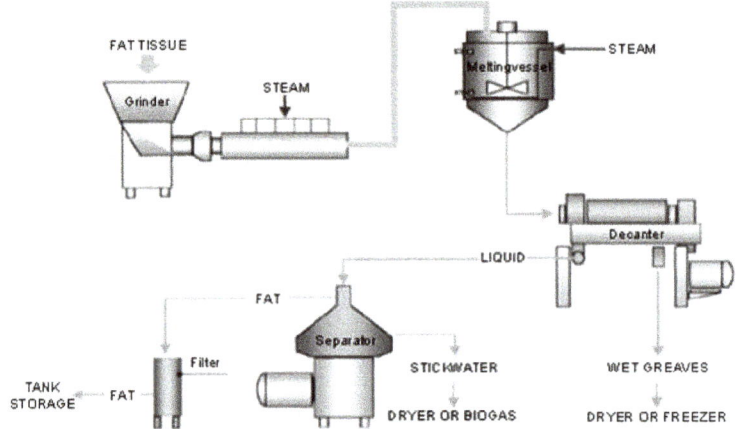

Figure 6. Wet rendering [59].

In dry rendering (Figure 7), in either batch or continuous processes, the fat tissues are cooked in their "own juices" with dry heat [59].

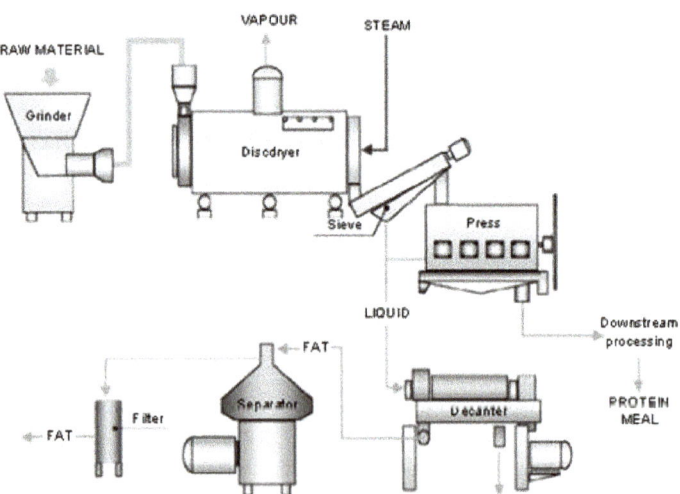

Figure 7. Dry rendering [60].

There is no rule when and where wet or dry rendering is ideal, but it can be observed that the lard and tallow from wet rendering are better than from dry rendering [59]. The rendering process may also be done using an organic solvent.

Usually, fats are further used in food, pet foods, feed applications but can also be transformed into soaps and oleochemicals (Figure 8), depending on the risk category [61]. All fats can be employed as feedstocks in biodiesel production.

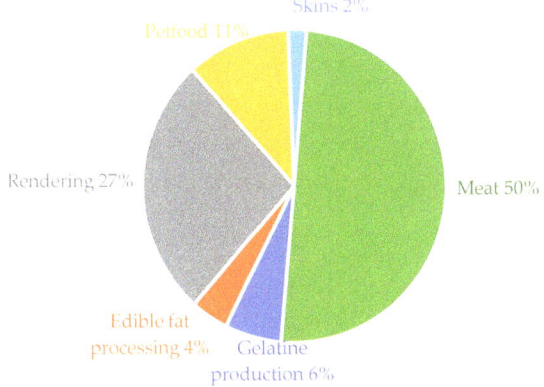

Figure 8. Estimated utilization of slaughtered animal (by % weight) [60].

2.6. Biodiesel Production from Animal Fats Versus Vegetable Oils

In terms of emission of pollutant gases, the advantages of replacing diesel with biodiesel produced from animal fats or obtained from vegetable oils are similar, since the emissions from burning generate similar results. However, Wyatt et al. have reported that three animal fats-based B20 biodiesel blends obtained from lard, beef tallow and chicken fat had lower nitrogen oxide (NOx) emission levels than B20 blend produced from soy oil [40].

In comparison with biodiesel from vegetable origin, biodiesel from animal fats has several advantages. Due to its lower content of unsaturated fatty acids, biodiesel produced from animal fats has a higher cetane number than biodiesel from the most vegetable oils and diesel fuel [18,36]. The cetane number increases with the increment of fatty acid carbon chains and the increase in degree of saturation [62]. A higher cetane number is recognized to lower NOx emissions [36]. Biofuel from animal fats has also a higher calorific value [36]. A nonconsensual issue is the oxidative stability of animal-based biodiesel. Some authors claim that animal fat-based biodiesel is less stable for oxidation due to the absence of natural oxidants as compared to biodiesel from vegetable oil [63–65]. On the other hand, others, claim that from the content of saturated fatty acid, the addition of animal fat improves the oxidative stability of biofuel [36,66,67]. Feedstocks rich in polyunsaturated fatty acids are more susceptible to oxidation, due to the presence of double bonds in the chains, than feedstocks rich in saturated or monounsaturated fatty acids [68].

Pereira et al. evaluated the effect of blending vegetable with animal-based biodiesel on the oxidative degradation of this biofuel. The authors reported that blends of soybean/beef tallow biodiesel presented a higher oxidative stabilities in comparison with soybean biodiesel [67]. Wyatt et al. also reported that the oxidative stability of biofuel from lard, beef tallow, and chicken fat is equivalent or better than soybean biodiesel [40]. However, Sendzikiene et al. [69] showed that biofuel from animal fats such as lard and tallow is less stable for oxidation than rapeseed and linseed oil. Fuel produced from fats has also some disadvantages, such as the higher cold filter plugging point (CFPP) due to a significant content of saturated fatty acids [63–65]. The CFPP refers to the lowest temperature at which a given volume of liquid fuel will still flow through a specific filter in a specified time when cooled under certain conditions [70]. This is an important property to cold temperature countries.

2.7. Catalysts for Biodiesel Production

In order to increase the reaction rate, the transesterification reaction needs to be catalyzed [20].

The catalyst is a substance that increases the reaction rate without being consumed. If the catalysis acts in the same phase as the reaction mixture is a homogeneous catalyst. However, if the catalysts acts in different phase it is classified as a heterogeneous catalyst [26]. In this case, the chemical reaction occurs at the interface between the two phases [71]. Figure 9 shows the different types of catalysts that can be used in the transesterification process.

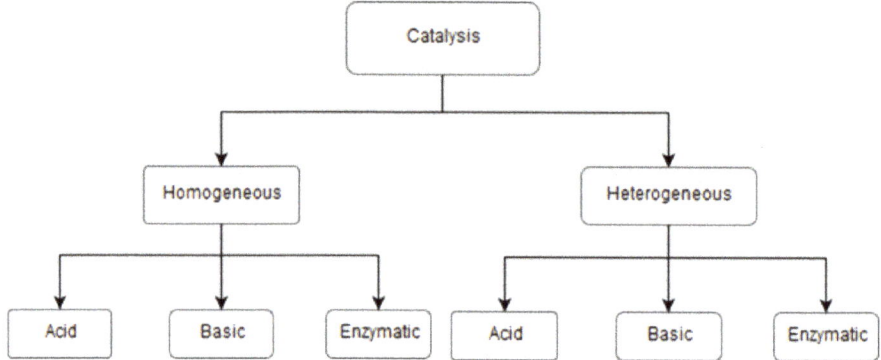

Figure 9. Different types of catalysis employed in the transesterification process.

Homogeneous catalysts have the main advantage of acting in the same phase of the reaction mixture, so the mass transfer resistance is minimized. Depending or their nature they can be basic, acid or enzymatic. These catalysts requires less time for a higher yield and conversion than the heterogeneous one [72].

Currently, the most common catalysts employed in the biodiesel industry are the homogeneous basic ones, such as sodium hydroxide (NaOH) and potassium hydroxide (KOH) that are easily soluble in methanol [22]. Homogeneous basic catalysts, having a higher reaction rate than homogenous acid ones, have the advantage of high biodiesel yield achieved in short reaction time under mild operating conditions. However, high purity feedstocks are essential and such catalytic systems should not be used with low grade fats feedstock which contains a high concentration of FFA and moisture. The FFA reacts with the basic catalyst forming soaps (Scheme 2), which leads to the losses of catalyst and reduced the biodiesel yields.

$$
\begin{array}{c}
\text{CH}_2\text{-O-CO-R}_1 \\
|\\
\text{CH-O-CO-R}_2 \\
|\\
\text{CH}_2\text{-O-CO-R}_3
\end{array}
+ \; 3\,\text{K-OH} \; \underset{}{\overset{H_2O}{\rightleftharpoons}} \; 3\,\text{R-CO-O}^-\text{-K}^+ \; + \;
\begin{array}{c}
\text{CH}_2\text{-OH} \\
|\\
\text{CH-OH} \\
|\\
\text{CH}_2\text{-OH}
\end{array}
$$

Triglycerides   Basic Catalyst   Soap   Glycerol

$$
R_1\text{-O-CO-H} \; + \; \text{K-OH} \; \xrightarrow{H_2O} \; R_1\text{-CO-O}^-\text{-K}^+ \; + \; H_2O
$$

FFA   Basic Catalyst   Soap   Water

Scheme 2. Saponification reaction of triglycerides and Neutralization reaction of FFA (adapted from [26]).

The feedstocks moisture, and the water formed in the above reaction (Scheme 2), can also hydrolyze the triglycerides into diglycerides and FFA, increasing the acidity index and decreasing the biodiesel yields, according to the reaction in Scheme 3.

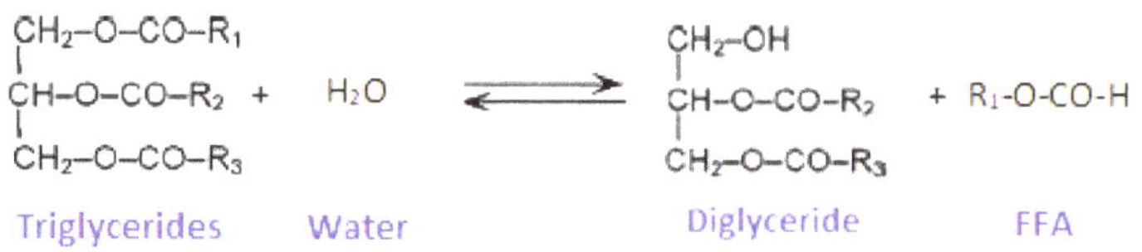

$$\begin{array}{c} CH_2-O-CO-R_1 \\ | \\ CH-O-CO-R_2 \\ | \\ CH_2-O-CO-R_3 \end{array} + H_2O \rightleftharpoons \begin{array}{c} CH_2-OH \\ | \\ CH-O-CO-R_2 \\ | \\ CH_2-O-CO-R_3 \end{array} + R_1-O-CO-H$$

Triglycerides   Water   Diglyceride   FFA

Scheme 3. Hydrolysis of triglycerides in FFA (adapted from [26]).

In order to overcome this issue, the transesterification reaction can be done in two stages. First, FFAs are converted into esters through pretreatment of the feedstock with an acid catalyst, reducing the FFA level (Scheme 4), followed by alkali transesterification. Another possibility is the use of a homogeneous acid catalyst such as sulfuric acid (H_2SO_4) or hydrochloric acid (HCl). Acid catalysts are not affected by FFA or water content due to their simultaneously capacity to catalyze both transesterification and esterification reactions (Scheme 4). Although, the acid catalyzed reaction is slower and thus, severe reaction conditions are needed, such as high reaction temperature, high acid catalyst concentration, and high alcohol:oil molar ratio in comparison with basic catalysts.

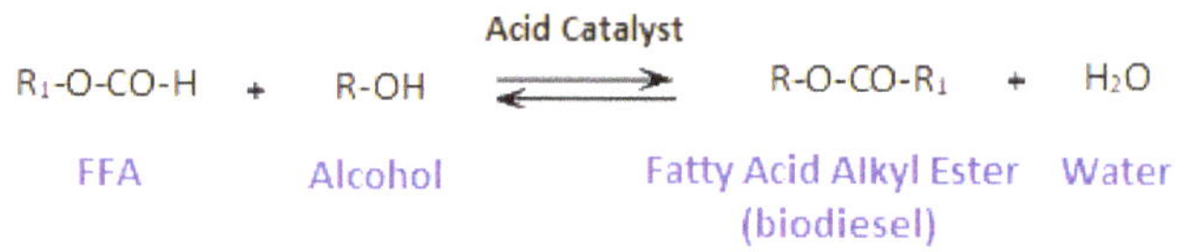

$$R_1-O-CO-H + R-OH \underset{}{\overset{Acid\ Catalyst}{\rightleftharpoons}} R-O-CO-R_1 + H_2O$$

FFA   Alcohol   Fatty Acid Alkyl Ester   Water
                                        (biodiesel)

Scheme 4. Esterification reaction of FFA (adapted from [26]).

Homogeneous catalysts cannot be separated from the reaction mixtures so they cannot be reused or regenerated, which is their major drawback. Besides that, these are corrosive to reactors and their separation from the reaction mixture is more difficult, requiring more complex equipment [73,74]. In the homogeneous catalysis purification process, a large amount of water is needed to neutralize and purify the biodiesel, producing a large quantity of wastewater, and increasing the production costs.

In the last decades, there has been a growing interest in the development and employment of new heterogeneous catalysts for biodiesel production. Heterogeneous or solid catalysts can be easily recovered, regenerated and reused. Depending or their nature they can be basic like alkaline earth metal oxides (CaO, MgO), hydrotalcite, acids like zirconia and alumina-based catalysts or enzymatic, like immobilized lipase [74,75]. Heterogeneous catalysts facilitate continuous reactor operation as they are easily separated from the reaction medium. They also minimize biodiesel and glycerin purification steps. Water consumption decreases because no catalyst neutralization process, and consequent steps, are required [21,74]. Like the homogeneous basic catalysts, the performance of basic solid catalysts is also affected by high FFA and water content, is also more active than solid acid one with mild operating conditions requirements. Mass transfer resistance is an issue when using heterogeneous catalysts due to the presence of three phases (oil/alcohol/catalyst) in the reaction mixture. In comparison with a homogeneous catalyst, solid catalyst presents lower conversions requiring more severe reaction conditions to achieve the same conversions [21,74]. Another issue to consider is the leaching of the active phase into the reaction mixture. The catalyst leaching leads to a homogeneous contribution. The extent of the catalyst leaching affects not only the life expectancy of catalysts and consequently their reuse but also the biodiesel quality [76]. For these reasons the heterogeneous catalyst should not be leaching and must be reused.

Lipases are enzymes that can be used in biodiesel production as catalysts, belonging to the homogeneous if used in its free form or heterogeneous group if immobilized [75]. Compared with other catalysts, enzymes have high selectivity, the products achieved (biodiesel and glycerin) are purer and no soaps are formed. Like acid catalysis (homogeneous or heterogeneous), enzymes have the capacity to catalyzed both triglycerides by transesterification and FFA by the esterification reaction. The main disadvantages are its high costs and the risk of enzyme inactivation by the short chain alcohols and products [75,77]. A literature survey on advantages and disadvantages of both homogeneous and heterogeneous catalysis is presented in Table 5.

Table 5. Advantages and disadvantages of homogeneous and heterogeneous catalysts in biodiesel production.

	Advantages	Disadvantages
Homogeneous	Less time required with a higher yield Mild operation condition Base catalysts are more active than acid Acid catalysts are not affected by the content of FFA or water simultaneously capacity to catalyzed transesterification and esterification reaction	Not separable from the reaction mixture, cannot be reuse A large amount of water is needed to neutralize and purify the biodiesel Base catalysts are affected by high FFA and water content
Heterogeneous	Easily recovered, regenerated and reused Available to batch or continuous fixed bed reactors Requires fewer process units with a simpler separation and purification processes The amount of water is reduced Base catalysts are more active than acid Acid catalysts are not affected by the FFA or water amount, capacity to catalyzed both transesterification and esterification reaction	Lower conversions requiring more severe reaction conditions to achieve the same conversions of homogeneous ones Mass transfer resistance due to the presence of three phases (oil/alcohol/catalyst) in the reaction mixture Base catalysts are affected by high FFA and water content

2.8. Homogeneous and Heterogeneous Catalytic Conversion of Animal Fats

Animal fats can be used as feedstocks to produces biodiesel through homogenously- (Table 6) and heterogeneously- (Table 7) catalyzed processes. Tables 6 and 7 summarize the main characteristics of both types of catalyzed processes.

2.9. Biodiesel Purification

After the transesterification reaction, biodiesel must be purified in order to accomplish the quality specifications (ASTM D6751 or EN 14214) and for further commercialization. In heterogeneous catalysis, in the first step, the solid catalyst can be easily recovered from the reaction mixture by filtration, instead, homogeneous ones will be dispersed and cannot be reused.

Table 6. Review of homogeneous catalysis to process animal fats.

Feedstock	Alcohol Type	Catalyst/ Co-Solvent	Reaction Conditions					Optimized Condition					Y = Yield; C = Conversion; P = Purity (%)	Ref.
			Reaction	Alcohol:Oil (Molar Ratio)	Cat. Conc. (wt % Oil)	Temp. (°C) and Impeller Speed	Time (h)	Catalyst	Alcohol:Oil (Molar Ratio)	Cat. Conc. (wt % Oil)	Temp. (°C) and Impeller Speed	Time (h)		
Mix of WFO and pork lard (fat fraction of mix 0–1(w/w))	MeOH	NaOH	-	6:1	0.8	60	1	-	6:1	0.8	60	1	Y = 81.7–87.7 P = 93.9–96.3	[57]
beef tallow	MeOH	KOH	-	6:1	1	60 and 60	2	-	6:1	1	60 and 60	2	Y = 90.6	[78]
pork lard													Y = 91.4	
chicken fat													Y = 76.8	
sardine oil													Y = 89.5	
corn oil		KOH					2	KOH	6:1	1		2	Y = 91	[79]
chicken fat		1st PT: H_2SO_4	-	-	-	60 and 300 rpm	1	1st PT: H_2SO_4	45 kg oil/40.5 kg alcohol	2.4 kg	60 and 300 rpm	1	Y (experiment 1) = 80.4; Y (experiment 2) = 81.5	
	MeOH	2nd PT: H_2SO_4					1	2nd PT: H_2SO_4	40 kg fat/3.1 kg alcohol	0.1 kg		1		
		KOH					2	KOH	6:1	1		2		
fleshing oil		1st PT: H_2SO_4					1	1st PT: H_2SO_4	50 kg oil/21.1 kg alcohol	0.62 kg		1	Y (experiment 1) = 81.6; Y (experiment 2) = 82.3	
		KOH					2	KOH	7.5:1	1		2		
beef tallow	MeOH	methanolic KOH	-	-	-	-	-	-	-	-	70–90	65 min.	Y = 96.26	[80]
beef tallow, pork lard	MeOH	KOH	-	6:1	0.8	60 and 60	2	-	6:1	0.8	60 and 60	2	Y = 90.8 Y = 91.5	[81]
chicken fat													Y = 76.8	

Table 6. Cont.

Feedstock	Alcohol Type	Catalyst/ Co-Solvent	Reaction Conditions					Optimized Condition					Y = Yield; C = Conversion; P = Purity (%)	Ref.
			Reaction	Alcohol:Oil (Molar Ratio)	Cat. Conc. (wt % Oil)	Temp. (°C) and Impeller Speed	Time (h)	Catalyst	Alcohol:Oil (Molar Ratio)	Cat. Conc. (wt % Oil)	Temp. (°C) and Impeller Speed	Time (h)		
Mix of waste fish oil(WFO); palm oil (PO) and WFO	MeOH	1st: H$_2$SO$_4$	-	6:1	1	60	1	-	6:1	1	60	1	FAME content (33.3 wt % PO; 66.7 wt % WFO) = 80	[82]
		2nd: NaOH	-	9:1	0.5	60	1	-	9:1	0.5	60	1	Y (33.3 wt % WFO; 66.7 wt % PO) = 98.5	
pork lard blended with n-hexane solvent	MeOH	KOH	-	6:1–18:1	0.48–3.05	50–60	2	-	10:1	2.0	60	2	Y (65 wt % solvent) = 98.2	[83]
Cyprinus carpio fish oil	MeOH	KOH; CH$_3$ONa; NaOH; EtONa/n-hexane; pet. ether; acetone; cyclohexane; diethyl ether	-	-	-	-	-	KOH	5:1	0.6	50	0.5	Y (1.5:1 hexane to methanol volume ratio) = 98.55 ± 1.02	[84]
beef tallow	MeOH	KOH	-	6:1	1.5	65 and 400 rpm	180 min	-	6:1	1.5	65 and 400 rpm	180 min	ester content = 95–97	[45]
lard	MeOH	NaOH	-	180 cm^3 fat/138 cm^3 MeOH	1.4 g	40–70	1.5	-	180 cm^3 fat/138 cm^3 MeOH	1.4 g	70	1.5	Y = 73	[85]
lard	MeOH	KOH	-	3.48–8.52	0.16–1.84	24.8–75.2	20 min.	-	7.5:1	1.26	65	20 min.	Y = 97.8 ± 0.6	[86]
Silurus triostegus heckel fish oil	MeOH	KOH	single and two-step transesterification	3:1–12:1	0.25–1.0	32–60	0.5–2	-	6:1	0.50	32	0.5	Y KOH = 96	[87]
		NaOH		6:1	0.5	32–60	1							
chicken fat	MeOH	KOH NaOH KOMe (32% in MeOH) NaOMe (30% in MeOH)	Sulfuric acid, hydrochloric acid and sulfamic acid with methanol	6:1	-	25–60	1–4	K methoxide	6:1	1	60	1	Y KOMe = 88.5	[88]

Table 6. Cont.

Feedstock	Alcohol Type	Catalyst/ Co-Solvent	Reaction	Reaction Conditions					Optimized Condition					Y = Yield; C = Conversion; P = Purity (%)	Ref.
				Alcohol:Oil (Molar Ratio)	Cat. Conc. (wt % Oil)	Temp. (°C) and Impeller Speed	Time (h)	Catalyst	Alcohol:Oil (Molar Ratio)	Cat. Conc. (wt % Oil)	Temp. (°C) and Impeller Speed	Time (h)		
chicken fat	MeOH	KOH	pre-treatment sulfuric acid and methanol	4:1–8:1	0.75–1.25	45	3–9 min	-	7:1	1	45	9 min	Y = 94.8	[89]
beef tallow	MeOH	NaOH	radio frequency heating	5:1–9:1	0.2–0.6	-	RF heating 1–5 min	-	9:1	0.6	-	RF heating 5 min	Y = 96.3 ± 0.5	[94]
tallow	MeOH	KOH	-	6:1	0.8	60	2	-	6:1	0.8	60	2	Y = 90.8	[91]
lard													Y = 91.4	
poultry													Y = 76.8	
mutton tallow	MeOH	KOH	pre-treatment sulfuric acid and MeOH	6:1	0.35–0.4 g	60 and 900 rpm	1.5	-	6:1	0.39 g	60 and 900 rpm	1.5	Y = 93.2	[92]
mix of chicken and swine fat residues	EtOH	KOH	animal fats pre-treated	6:1–8:1	0.44–1.32	30–70	1	-	7:1	0.96	30	1	C = 83.5	[95]

PT = pretreatment.

Table 7. Review of heterogeneous catalysis to processes animal fats.

Feedstock	Alcohol Type	Catalyst/ Co-Solvent	Reaction Conditions							Optimized Condition					Ref.
			Catalyst Preparation	Alcohol:Oil (Molar Ratio)	Cat. Conc. (wt % Oil)	Temp. (°C) and Impeller Speed	Time (h)	Catalyst Preparation	Alcohol:Oil (Molar Ratio)	Cat. Conc. (wt % Oil)	Temp. (°C and Impeller Speed)	Time (h)	Y = Yield; C = Conversion; P = Purity (%)	
catfish fat (*Pangasius*)	MeOH	barnacle	900 °C	6:1–15:1	2–7	65	2–8	900	12:1	5	65	4	FAME content = 97.2 ± 0.04 wt %	[49]
		bivalve clam											FAME content = 96.9 ± 0.03 wt %	
waste lard from piglet roasting (compared to heated and unheated lard)	MeOH	quicklime and CaO	Pure CaO 550 °C, 2 h; Quicklime 550 °C, 4 h	6:1	5	40; 50; 60 and 900 rpm	up to 3	Quicklime 550 °C 4 h	6:1	5	60; 900 rpm	1	FAME concentration waste lard = 97.5%	[94]
chicken fat	MeOH	crab, cockle shells and mix	900 °C, 2 h	-	-	-	-	900 °C, 2 h	13.8:1	4.9% (1:1 crab:cockle)	65	3	P = 98.8	[48]
commercial-grade fat	MeOH	Amberlyst™ A26 OH	-	6:1	1–2.7 mol/L of fat	65	60–500 min	-	6:1	2.2 mol/L	65	6	Y = 90–95	[95]
pork lard	MeOH	CaMnOx	-	9:1–27:1	1.5	40–60	4–8	-	21:1	1	60	8	Y = 99.6	[64]
soybean oil (SBO) and beef tallow (bf)	MeOH	Sulfonated polystyrene compounds	-	3:1–9:1; 100:1	20 mol % of –SO$_3$H groups in the oil mass	28–64	3–18	-	100:1	20 mol % of –SO$_3$H groups in relation to the oil mass	64	18	C$_{SBO}$ = 85	[96]
	EtOH		-	100:1		64	18	-	-	-	-	-	C$_{Beef\ tallow}$ = 85 and 75	
beef tallow	MeOH	K^7/CaO-Fe$_3$O$_4$	-	3:1–12:1	1–6 g	40–65	20–70 min.	-	10:1	5 g	55	1	Y = 94	[97]
beef tallow	MeOH	Cs$_2$O/γ-Al$_2$O$_3$	wet impregnation with aqueous solution of Cs$_2$CO$_3$.	8:1–12:1	4–6 g	55–75	80–160 min.	-	10.5:1	5.3%	66	2	Y = 95.5	[98]

Biodiesel phase can be separated from the glycerin phase in a decanter by sedimentation or centrifugation due to their density difference [99,100]. The excess alcohol used in the reaction can be separated from both phases by evaporation or distillation. The obtained biodiesel still contains free glycerol, soap, residual alcohol, catalyst and mono, di, and triglycerides unreacted, which must be removed. Table 8 summarizes the main consequences of biodiesel contamination on internal combustion engines.

Table 8. Biodiesel impurities and properties effects on internal combustion engines [101,102].

Impurities	Effects	
	Biodiesel	Engines
FFA	Low oxidation stability	Corrosion
water	Reduces the heat of combustion Hydrolysis (FFA production)	Corrosion Bacteriological growth (filter blockage)
methanol	Low values of density and viscosity Low flash point (transport, storage, and use problems)	Corrosion of Al and Zn pieces
glycerides	High viscosity	Deposits in the injectors (carbon residue) Crystallization
metals (soap, catalyst)	-	Deposits in the injectors (carbon residue) Filter blockage (sulfated ashes) Engine weakening
glycerin	Decantation storage problem	Increase aldehydes and acrolein emissions

Biodiesel can be purified by several processes. The most used industrial biodiesel purification process is water washing. This method is simple, efficient and allows one to obtain biodiesel with high purity. Distilled water, deionized water, acidulated water, acid and water or water, and an organic solvent can be added to biodiesel [99,102,103]. Due to their water solubility, residual glycerol, methanol, catalyst, and any soap formed during the reaction can be eliminated. Lastly, washed biodiesel need to be dehydrated [99,103]. Biodiesel purification with water is time-consuming and produces large volumes of wastewater which cannot be discharged in watercourses. Wastewater effluent has to be treated, increasing the biodiesel production costs. To overcome this issue, a dry washing purification process, using solid sorbents such as ion exchange resin, silica, activated carbon among other adsorbents, can be adopted. This purification process is fast, easy to integrate in an industrial plant and being water-free, no wastewater is produced lowering the production cost [102]. The main drawback is the purified biodiesel may not meet methanol and glycerin EN 14214 specifications. Therefore, methanol and glycerin should be removed as much as possible before the purification process [101,104]. Also, the adsorbent cost, its recycling, and elimination can be a disadvantage [105].

Inorganic and polymeric membranes can also be employed for biodiesel purification [104]. Membrane works as a selective barrier retaining the biodiesel contaminants [105]. No water is consumed within this process, consequently, no wastewater is produced although the addition of a small amount of water improves glycerin retention. High-quality biodiesel meeting the required specifications can be achieved. This process presents some issues too such as, membrane cleaning, membrane costs and the increase of biodiesel production cost [103].

2.10. Quality Specifications

The quality of biodiesel can be influenced by several factors that may be reflected in its chemical and physical properties. To be commercialized, biodiesel has to accomplish the quality specifications established by institutions like the European Committee of Standardization (ISO) and the American

Society for Testing and Materials (ASTM) (Table 9). These regulations, which are dynamic and must be periodically reviewed, describe not only the quality requirements but also the test methods employed.

Table 9. ASTM D6751 and EN 14214 specifications of biodiesel fuels (B 100) [106,107].

Property Specification	ASTM D6751 Limit	Test Methods	EN 14214 Limit	Test Methods
Ester content (% (m/m))	-	-	96.5	EN 14103
Density at 15 °C (kg/m^3)	880	D1298	860–900	EN ISO 3675/12185
Viscosity at 40 °C (mm^2/s)	1.9–6.0	D445	3.5–5.0	EN ISO 3104
Cetane number	Min. 47	D613	Min. 51.0	EN ISO 5165
Iodine number (g I$_2$/100 g)	-	-	Max. 120	EN 14111/16300
Acid value (mg KOH/g)	Max. 0.50	D664	Max. 0.50	EN 14104
Pour point (°C)	-15 to -16	D97	-	-
Flash point (°C)	Min. 130	D93	Min. 101	EN ISO 2719/3679
Cloud point (°C)	-3 to -12	D2500	-	-
Cold filter plugging point (°C)	Max. +5	D6371	-	EN 116/16329
Copper strip corrosion (3 h at 50 °C)	No 3	D130	class 1	EN ISO 2160
Carbon residue (% (m/m))	Max. 0.05	D4530	-	-
Methanol content (% (m/m))	Max. 0.20	EN 14110	Max. 0.20	EN 14110
Water content (mg/kg)	Max. 500	D2709	Max. 500	EN ISO 12937
Sulfur (mg/kg)	S15 Max. 15 S500 Max. 500	D5453	Max. 10.0	EN ISO 20846/20884
Sulfated ash (% (m/m))	Max. 0.02	D874	Max. 0.02	EN ISO 3987
Phosphorus content (mg/kg)	Max. 10	D4951	Max. 4.0	EN 14107/16294
Free glycerol (% (m/m))	Max. 0.02	D6584	Max. 0.02	EN 14105/EN 14106
Total glycerol (% (m/m))	Max. 0.24	D6548	Max. 0.25	EN 14105
Monoglyceride (% (m/m))	Max. 0.40	D6584	Max. 0.70	EN 14105
Diglyceride (% (m/m))	-	-	Max. 0.20	EN 14105
Triglyceride (% (m/m))	-	-	Max. 0.20	EN 14105
Distillation temperature, 90% recovered (°C)	Max. 360	D1160	-	-
Oxidation stability h (at 110 °C)	Min. 3	EN 15751	Min. 8	EN 14112/15751
Linolenic acid methyl ester (% (m/m))	-	-	Max. 12.0	EN 14103
Polyunsaturated (≥4 double bonds) Methyl esters (% (m/m))	-	-	Max. 1.00	EN 15779
Alkaline metals (Na$^+$ K) (mg/kg)	Max. 5.0	EN 14538	Max. 5.0	EN 14108/14109/14538
Alkaline earth metals (Ca + Mg) (mg/kg)	Max. 5.0	EN 14538	Max. 5.0	EN 14538
Total contamination (mg/kg)	-	-	Max. 24	EN 12662

2.11. Properties of Biodiesel from Different Feedstocks

Biodiesel properties are influenced not only by raw materials but also by transesterification and purity process. Table 10 shows a literature review of biodiesel properties from different feedstocks.

Table 10. Biodiesel properties from different feedstocks.

Property Specification	ASTM D6751 Limit	EN 14214 Limit	Rapeseed Oil [3]	Soybean Oil [3]	Palm Oil [3]	Chicken Fat [79]	Fleshing Oil [79]	Beef Tallow [3,18]	Mutton Tallow [18]	Lard [18]	Fish Oil [108]	WCO [108]
Density at 15 °C (kg/m^3)	880	860–900	882	914	864	867–889.7	875.5–876.7	832–872	856–882	873–877.4	881–890	875–888
Viscosity at 40 °C (mm^2/s)	1.9–6.0	3.5–5.0	4.44	4.04	4.5	4.94–6.25	4.7–4.77	4.89–5.35	4.75–5.98	4.59–5.08	3.82–7.2	3.66–6.8
Cetane number	Min. 47	Min. 51.0	54.4	37.9	54.6	52.3	58.8	60.36	59–59	-	50.9–52.6	41–66
Iodine number (g I$_2$/100 g)	-	Max. 120	-	128–143	54	95.5–130	53.6–61	nm–44.4	40–126	67–77	nm–185	60–125.21
Acid value (mg KOH/g)	Max. 0.50	Max. 0.50	-	0.266	0.24	0.22–0.8	0.28–0.32	0.147–0.2	0.3–0.65	0.04–1.13	0.35–1.32	0.27–1.31
Pour point (°C)	−15 to −16	-	−12	-	15	−6–12.3	-	10–15	−5	5–7	−14–4	−2.5–9
Flash point (°C)	Min. 130	Min. 101	170	254	135	169–174	168–175	152–171	-	143.5–147	114–176	70.6–190
Cloud point (°C)	−3 to −12	-	−3.3	0.9	16	−5–14	-	nm–16	−4	-	-	−12–13
Cold filter plugging point (°C)	Max. +5	-	−13	−4	12	2–3	10–11	nm–14	-	-	−5	−5–12
Copper strip corrosion (3 h at 50 °C)	No 3	class 1	-	1	1	No 1	No 1	No 1	-	No 1	-	-
Carbon residue (% (m/m))	Max. 0.05	-	81	-	-	nm–0.024	-	-	-	nm–0.21	76.53–80.01	0.0004–77.38
Methanol content (% (m/m))	Max. 0.20	Max. 0.20	-	-	-	0.01–0.06	0.01–0.01	nm–0.1	-	-	-	-
Water content (mg/kg)	Max. 500	Max. 500	-	<0.005 %vol	-	200–440	326–410	nm–374.2	-	184–1100	-	-
Sulfur (mg/kg)	S15 Max. 15 S500 Max. 500	Max. 10.0	-	0.8	0.003	nm–81.5	138.1–141	nm–7.0	-	-	-	0–12.5
Sulfated ash (% (m/m))	Max. 0.02	Max. 0.02	-	<0.005	0.002	-	0.03	nm–<0.005	nm–0.025	nm–0.002	-	-
Phosphorus content (mg/kg)	Max. 10	Max. 4.0	-	0.1	<0.001	-	100	nm–<0.1	nm–16	-	-	-
Free glycerol (% (m/m))	Max. 0.02	Max. 0.02	-	0.012	0.01	0.008–0.02	0.01–0.01	0.008–0.01	-	-	-	-

Table 10. Cont.

Property Specification	ASTM D6751 Limit	EN 14214 Limit	Rapeseed Oil [3]	Soybean Oil [3]	Palm Oil [3]	Chicken Fat [79]	Fleshing Oil [79]	Beef Tallow [3,18]	Mutton Tallow [18]	Lard [18]	Fish Oil [108]	WCO [108]
Total glycerol (% (m/m))	Max. 0.24	Max. 0.25	-	0.149	0.01	0.03–0.19	0.10–0.05	0.076–0.33	-	-	-	-
Monoglyceride (% (m/m))	Max. 0.40	Max. 0.70	0.473	-	-	0.02–0.56	0.06–0.27	0.13–0.223	-	-	-	-
Diglyceride (% (m/m))	-	Max. 0.20	0.088	-	-	0.05–0.09	0.02–0.09	0.63–0.12	-	-	-	-
Triglyceride (% (m/m))	-	Max. 0.20	0.019	-	-	0.06–0.12	0.04–0.20	0–0.07	-	-	-	-
Distillation temperature, 90 % recovered (°C)	Max. 360	-	-	-	-	-	-	307–344	-	nm–352.5	-	-
Oxidation stability h (at 110 °C)	Min. 3	Min. 8	7.6	2.1	10.3	nm–6	-	nm–1.6	-	0.9–1.4	-	0.43–15.9
Linolenic acid methyl ester (% (m/m))	-	Max. 12.0	-	-	-	-	-	-	-	-	-	-
Polyunsaturated (≥4 double bonds) Methyl esters (% (m/m))	-	Max. 1.00	-	-	-	-	-	-	-	-	-	-
Alkaline metals (Na$^+$ K) (mg/kg)	Max. 5.0	Max. 5.0	-	-	-	-	5	2–2.63	-	nm–17.2	-	-
Alkaline metals (Ca + Mg) (mg/kg)	Max. 5.0	Max. 5.0	-	-	-	-	-	-	-	-	-	-
Total contamination (mg/kg)	-	Max. 24	-	-	-	-	-	-	-	-	-	-
Heat of combustion (MJ/<g)	-	-	37	39.76	-	39.34–40.17	39.89–39.95	40.23	-	36.5–40.10	37.79–42.24	35.40–43.21

nm—not measured.

3. Conclusions

Biodiesel obtained by alcoholysis of fats is a feasible low carbon fuel to replace the conventional fossil diesel thus helping to mitigate the anthropogenic carbon emissions. First generation biodiesel, obtained by methanolysis of vegetable oils, presents severe sustainability issues related to the use of arable lands to produce energy-dedicated crops (oleaginous crops). Biodiesel sustainability issues can be minimized by using non-edible fats such as animal fats and waste cooking oils. Replacing methanol by ethanol could also contribute to reducing carbon emissions from biodiesel because ethanol can be obtained by biomass fermentation, thus being a renewable alcohol.

Biodiesel production processes can be improved by replacing conventional homogeneous (basic) catalysts with heterogeneous catalysts. Among the huge number of scientific papers on heterogeneous catalysts for biodiesel production, the excellent performances (catalytic activity) of calcium-based catalysts stands out but they have never been tested industrially. The lack of data on the stability of calcium catalysts appears to be a limitation to their industrial testing. Dry-washing purification of biodiesel, instead of the wet process nowadays in use, can also contribute to biodiesel sustainability. Biodiesel dry-washing decreases the large volumes of wastewater generated in the traditional purification method and cuts down the energy required in the biodiesel drying process.

Author Contributions: All authors participated equally in the conception of the presented review and the manuscript was prepared by M.R. and A.P.S.D. and revised by J.G., J.F.P. and J.C.B. contributed to the funding obtaining (PTDC/EMS-ENE/4865/2014).

Funding: The authors acknowledge FCT (Fundação para a Ciência e Tecnologia, Portugal) for funding project PTDC/EMS-ENE/4865/2014. This work was also supported through IDMEC, under LAETA, project FCT UID/EMS/50022/2019.

Conflicts of Interest: The authors declare no conflict of interest.

Abbreviations

ASTM	American Society for Testing and Materials
CFPP	Cold Filter Plugging Point
CIS	Commonwealth of Independent States
EIA	U.S. Energy Information Administration
ETBE	Ethyl *tert*-butyl ether
FAEE	Fatty acid ethyl ester
FAME	Fatty acid methyl ester
FFA	Free fatty acid
GHG	Greenhouse gas
HC	Hydrocarbons
ISO	European Committee of Standardization
MTBE	Methyl *tert*-butyl ether
NOx	Nitrous oxide
PM	Particulate matter
SRM	Specified Risk Material
TSE	Transmissible Spongiform Encephalopathies
UCO	Used cooking oil
WAF	Waste animal fats
WCO	Waste cooking oil

References

1. U.S. Energy Information Administration. *International Energy Outlook 2017*; U.S. Energy Information Administration: Washington, DC, USA, 2017.
2. *BP Statistical Review of World Energy 2019*, 68th ed.; BP: London, UK, 2019.

3. Atabani, A.E.; Silitonga, A.S.; Badruddin, I.A.; Mahlia, T.M.I.; Masjuki, H.H.; Mekhilef, S. A comprehensive review on biodiesel as an alternative energy resource and its characteristics. *Renew. Sustain. Energy Rev.* **2012**, *16*, 2070–2093. [CrossRef]
4. European Commission 2020 Climate & Energy Package. Available online: https://ec.europa.eu/clima/policies/strategies/2020_en?fbclid=IwAR3DhvhN-_IcDcqtIu0_sgWiB7cctQrqR90JKNs5dMRxOCxtGV_FQQVw4QM (accessed on 5 February 2019).
5. European Union Directive (EU) 2009/28. Available online: https://eur-lex.europa.eu/LexUriServ/LexUriServ.do?uri=OJ:L:2009:140:0016:0062:EN:PDF (accessed on 12 February 2019).
6. European Environment Agency (EEA). Use of Renewable Fuels in Transport. Available online: https://www.eea.europa.eu/data-and-maps/indicators/use-of-cleaner-and-alternative-fuels/use-of-cleaner-and-alternative-4 (accessed on 12 February 2019).
7. European Commission Directive (EU) 2015/1513. Available online: https://eur-lex.europa.eu/legal-content/EN/TXT/?uri=celex%3A32015L1513 (accessed on 13 February 2019).
8. European Commission 2030 Climate & Energy Framework. Available online: https://ec.europa.eu/clima/policies/strategies/2030_en (accessed on 11 March 2019).
9. European Commission. The Commission Presents Strategy for a Climate Neutral Europe by 2050—Questions and Answers. Available online: https://ec.europa.eu/clima/policies/strategies/2050_en (accessed on 13 March 2019).
10. European Commission Directive (EU) 2003/30. Available online: https://eur-lex.europa.eu/legal-content/EN/ALL/?uri=celex%3A32009L0028 (accessed on 13 March 2019).
11. Biodiesel Explained—U.S. Energy Information Administration (EIA). Available online: https://www.eia.gov/energyexplained/biofuels/biodiesel.php (accessed on 14 April 2019).
12. Phillips, S.; Flach, B.; Lieberz, S.; Lappin, J.; Bolla, S. EU Biofuels Annual 2018. Available online: https://apps.fas.usda.gov/newgainapi/api/report/downloadreportbyfilename?filename=Biofuels%20Annual_The%20Hague_EU-28_7-3-2018.pdf (accessed on 14 April 2019).
13. Direcção Geral de Energía e Geología. *Renováveis-Estatísticas Rápidas—No 169*; Direcção Geral de Energía e Geología: Lisbon, Portugal, 2018.
14. Mahmudul, H.M.; Hagos, F.Y.; Mamat, R.; Adam, A.A.; Ishak, W.F.W.; Alenezi, R. Production, characterization and performance of biodiesel as an alternative fuel in diesel engines—A review. *Renew. Sustain. Energy Rev.* **2017**, *72*, 497–509. [CrossRef]
15. Romano, S.D.; Sorichetti, P.A. *Dielectric Spectroscopy in Biodiesel Production and Characterization*; Springer: London, UK, 2011.
16. Islam, A.; Ravindra, P. *Biodiesel Production with Green Technologies*; Springer: Basel, Switzerland, 2016; ISBN 9783319452722.
17. Moser, B.R. Biodiesel production, properties, and feedstocks. *Vitr. Cell. Dev. Biol. Plant* **2009**, *45*, 229–266. [CrossRef]
18. Banković-Ilić, I.B.; Stojković, I.J.; Stamenković, O.S.; Veljkovic, V.B.; Hung, Y.T. Waste animal fats as feedstocks for biodiesel production. *Renew. Sustain. Energy Rev.* **2014**, *32*, 238–254. [CrossRef]
19. Banerjee, A.; Chakraborty, R. Parametric sensitivity in transesterification of waste cooking oil for biodiesel production—A review. *Resour. Conserv. Recycl.* **2009**, *53*, 490–497. [CrossRef]
20. Shahid, E.M.; Jamal, Y. Production of biodiesel: A technical review. *Renew. Sustain. Energy Rev.* **2011**, *15*, 4732–4745. [CrossRef]
21. Avhad, M.R.; Marchetti, J.M. Innovation in solid heterogeneous catalysis for the generation of economically viable and ecofriendly biodiesel: A review. *Catal. Rev.* **2016**, *58*, 157–208. [CrossRef]
22. Ajala, O.E.; Aberuagba, F.; Odetoye, T.E.; Ajala, A.M. Biodiesel: Sustainable energy replacement to petroleum-based diesel fuel—A review. *Chembioeng Rev.* **2015**, *2*, 145–156. [CrossRef]
23. Mathiyazhagan, M.; Ganapathi, A. Factors affecting biodiesel production. *Res. Plant Biol.* **2011**, *1*, 1–5.
24. Tan, H.W.; Abdul Aziz, A.R.; Aroua, M.K. Glycerol production and its applications as a raw material: A review. *Renew. Sustain. Energy Rev.* **2013**, *27*, 118–127. [CrossRef]
25. Luo, X.; Ge, X.; Cui, S.; Li, Y. Value-added processing of crude glycerol into chemicals and polymers. *Bioresour. Technol.* **2016**, *215*, 144–154. [CrossRef] [PubMed]

26. Borges, M.E.; Díaz, L. Recent developments on heterogeneous catalysts for biodiesel production by oil esterification and transesterification reactions: A review. *Renew. Sustain. Energy Rev.* **2012**, *16*, 2839–2849. [CrossRef]
27. Avramović, J.M.; Veličković, A.V.; Stamenković, O.S.; Rajković, K.M.; Milić, P.S.; Veljković, V.B. Optimization of sunflower oil ethanolysis catalyzed by calcium oxide: RSM versus ANN-GA. *Energy Convers. Manag.* **2015**, *105*, 1149–1156. [CrossRef]
28. Dalena, F.; Senatore, A.; Marino, A.; Gordano, A.; Basile, M.; Basile, A. Methanol production and applications: An overview. In *Methanol: Science and Engineering*; Elsevier: Amsterdam, The Netherlands, 2017; pp. 3–28.
29. Sarkar, N.; Ghosh, S.K.; Bannerjee, S.; Aikat, K. Bioethanol production from agricultural wastes: An overview. *Renew. Energy* **2012**, *37*, 19–27. [CrossRef]
30. Watcharathamrongkul, K. Calcium oxide based catalysts for ethanolysis of soybean oil. *Songklanakarin J. Sci. Technol.* **2010**, *32*, 627–634.
31. Velickovic, A.; Avramovic, J.; Stamenkovic, O.; Veljkovic, V. Kinetics of the sunflower oil ethanolysis using CaO as catalyst. *Chem. Ind. Chem. Eng. Q.* **2016**, *22*, 409–418. [CrossRef]
32. Rashid, U.; Ibrahim, M.; Ali, S.; Adil, M.; Hina, S.; Bukhari, I.H.; Yunus, R. Comparative study of the methanolysis and ethanolysis of Maize oil using alkaline catalysts. *Grasas Aceites* **2012**, *63*, 35–43. [CrossRef]
33. Verma, P.; Sharma, M.P. Comparative analysis of effect of methanol and ethanol on Karanja biodiesel production and its optimisation. *Fuel* **2016**, *180*, 164–174. [CrossRef]
34. García, M.; Gonzalo, A.; Sánchez, J.L.; Arauzo, J.; Simoes, C. Metanolysis and ethanolysis of animal fats: A comparative study of the influence of alcohols. *Chem. Ind. Chem. Eng. Q.* **2011**, *17*, 91–97. [CrossRef]
35. Meneghetti, S.M.P.; Meneghetti, M.R.; Wolf, C.R.; Silva, E.C.; Lima, G.E.S.; de Lira Silva, L.; Serra, T.M.; Cauduro, F.; de Oliveira, L.G. Biodiesel from castor oil: A comparison of ethanolysis versus methanolysis. *Energy Fuels* **2006**, *20*, 2262–2265. [CrossRef]
36. Adewale, P.; Dumont, M.J.; Ngadi, M. Recent trends of biodiesel production from animal fat wastes and associated production techniques. *Renew. Sustain. Energy Rev.* **2015**, *45*, 574–588. [CrossRef]
37. Fat Minami-Nutrition. Available online: https://socratic.org/questions/how-do-lipids-affect-the-digestive-systemhttp://www.minami-nutrition.co.uk/website/absorption.php (accessed on August 2006).
38. Woodgate, S.L.; Van der Veen, J.T. Fats and Oils—Animal Based. In *Food Processing: Principles and Applications*, 2nd ed.; Clark, S., Jung, S., Lamsal, B., Eds.; John Wiley & Sons.: Hoboken, NJ, USA, 2014; pp. 481–499.
39. Azócar, L.; Ciudad, G.; Heipieper, H.J.; Muñoz, R.; Navia, R. Improving fatty acid methyl ester production yield in a lipase-catalyzed process using waste frying oils as feedstock. *J. Biosci. Bioeng.* **2010**, *109*, 609–614. [CrossRef] [PubMed]
40. Wyatt, V.T.; Hess, M.A.; Dunn, R.O.; Foglia, T.A.; Haas, M.J.; Marmer, W.N. Fuel properties and nitrogen oxide emission levels of biodiesel produced from animal fats. *J. Amer Oil Chem Soc.* **2005**, *82*, 585–591. [CrossRef]
41. Dias, J.M.; Alvim-ferraz, M.C.M.; Almeida, M.F. Mixtures of vegetable oils and animal fat for biodiesel production: Influence on product composition and quality. *Energy Fuels* **2008**, *20*, 3889–3893. [CrossRef]
42. Leung, D.Y.C.; Guo, Y. Transesterification of neat and used frying oil: Optimization for biodiesel production. *Fuel Process. Technol.* **2006**, *87*, 883–890. [CrossRef]
43. Charoenchaitrakool, M.; Thienmethangkoon, J. Statistical optimization for biodiesel production from waste frying oil through two-step catalyzed process. *Fuel Process. Technol.* **2011**, *92*, 112–118. [CrossRef]
44. Chakraborty, R.; Gupta, A.K.; Chowdhury, R. Conversion of slaughterhouse and poultry farm animal fats and wastes to biodiesel: Parametric sensitivity and fuel quality assessment. *Renew. Sustain. Energy Rev.* **2014**, *29*, 120–134. [CrossRef]
45. Espinosa, M.; Canielas, L.; Silvana, M.; Moraes, A.; Schmitt, C.; Assis, R.; Rodrigues, S.; Regina, M.; Rodrigues, A.; Bastos, E. Beef tallow biodiesel produced in a pilot scale. *Fuel Process. Technol.* **2009**, *90*, 570–575.
46. Ito, T.; Sakurai, Y.; Kakuta, Y.; Sugano, M.; Hirano, K. Biodiesel production from waste animal fats using pyrolysis method. *Fuel Process. Technol.* **2012**, *94*, 47–52. [CrossRef]
47. Dias, J.M.; Ferraz, C.A.; Almeida, M.F. Using mixtures of waste frying oil and pork lard to produce biodiesel. *Energy Fuels* **2008**, *22*, 3889–3893.
48. Boey, P.L.; Maniam, G.P.; Hamid, S.A.; Ali, D.M.H. Crab and cockle shells as catalysts for the preparation of methyl esters from low free fatty acid chicken fat. *J. Am. Oil Chem. Soc.* **2011**, *88*, 283–288. [CrossRef]

49. Maniam, G.P.; Hindryawati, N.; Nurfitri, I.; Manaf, I.S.A.; Ramachandran, N.; Rahim, M.H.A. Utilization of waste fat from catfish (Pangasius) in methyl esters preparation using CaO derived from waste marine barnacle and bivalve clam as solid catalysts. *J. Taiwan Inst. Chem. Eng.* **2015**, *49*, 58–66. [CrossRef]
50. Aryee, A.N.A.; Simpson, B.K.; Cue, R.I.; Phillip, L.E. Enzymatic transesterification of fats and oils from animal discards to fatty acid ethyl esters for potential fuel use. *Biomass Bioenergy* **2011**, *35*, 4149–4157. [CrossRef]
51. Behçet, R. Performance and emission study of waste anchovy fish biodiesel in a diesel engine. *Fuel Process. Technol.* **2011**, *92*, 1187–1194. [CrossRef]
52. Knothe, G.; Razon, L.F. Biodiesel fuels. *Prog. Energy Combust. Sci.* **2017**, *58*, 36–59. [CrossRef]
53. Živković, S.B.; Veljković, M.V.; Banković-Ilić, I.B.; Krstić, I.M.; Konstantinović, S.S.; Ilić, S.B.; Avramović, J.M.; Stamenković, O.S.; Veljković, V.B. Technological, technical, economic, environmental, social, human health risk, toxicological and policy considerations of biodiesel production and use. *Renew. Sustain. Energy Rev.* **2017**, *79*, 222–247. [CrossRef]
54. Canakci, M. The potential of restaurant waste lipids as biodiesel feedstocks. *Bioresour. Technol.* **2007**, *98*, 183–190. [CrossRef]
55. Index Mundi Commodity Prices. Available online: https://www.indexmundi.com/commodities/ (accessed on 5 February 2019).
56. Grenea WASTE-BASED MARKET PERFORMANCE. Available online: https://www.greenea.com/en/market-analysis/ (accessed on 5 February 2019).
57. EFPRA. What are the Three Categories? Available online: http://efpra.eu/which-byproducts-rendered/ (accessed on 22 November 2018).
58. Meeker, D.L. *Essential Rendering, All About The Animal By-Products Industry*; The National Renderers Association: Alexandria, VA, USA, 2006; ISBN 0965466035.
59. Jayathilakan, K.; Sultana, K.; Radhakrishna, K.; Bawa, A.S. Utilization of byproducts and waste materials from meat, poultry and fish processing industries: A review. *J. Food Sci. Technol.* **2012**, *49*, 278–293. [CrossRef]
60. Woodgate, S.; van der Veen, J. The role of fat processing and rendering in the European Union animal production industry. *Biotechnol. Agron. Soc. Environ.* **2004**, *8*, 283–294.
61. Giriprasad, R.; Sharma, H.; Goswami, M. Animal fat-processing and its quality control. *J. Food Process. Technol.* **2013**, *4*, 252. [CrossRef]
62. Knothe, G.; Krahl, J.; Van Gerpen, J. *The Biodiesel Handbook*; AOCS Press: Urbana, IL, USA, 2005; Volume 1.
63. Encinar, J.M.; Sánchez, N.; Martínez, G.; García, L. Study of biodiesel production from animal fats with high free fatty acid content. *Bioresour. Technol.* **2011**, *102*, 10907–10914. [CrossRef] [PubMed]
64. Dias, J.M.; Alvim-Ferraz, M.C.M.; Almeida, M.F.; Méndez Díaz, J.D.; Polo, M.S.; Utrilla, J.R. Selection of heterogeneous catalysts for biodiesel production from animal fat. *Fuel* **2012**, *94*, 418–425. [CrossRef]
65. Avhad, M.R.; Marchetti, J.M. A review on recent advancement in catalytic materials for biodiesel production. *Renew. Sustain. Energy Rev.* **2015**, *50*, 696–718. [CrossRef]
66. Pinzi, S.; Leiva-Candia, D.; López-García, I.; Redel-Macías, M.; Dorado, M. Latest trends in feedstocks for bioidesel production. *Biofuels Bioprod. Biorefin.* **2014**, *8*, 126–143. [CrossRef]
67. Pereira, G.G.; Garcia, R.K.A.; Ferreira, L.L.; Barrera-Arellano, D. Soybean and soybean/beef-tallow biodiesel: A comparative study on oxidative degradation during long-term storage. *J. Am. Oil Chem. Soc.* **2017**, *94*, 587–593. [CrossRef]
68. Canakci, M.; Sanli, H. Biodiesel production from various feedstocks and their effects on the fuel properties. *J. Ind. Microbiol. Biotechnol.* **2008**, *35*, 431–441. [CrossRef]
69. Sendzikiene, E.; Makareviciene, V.; Janulis, P. Oxidation stability of biodiesel fuel produced from fatty wastes. *Pol. J. Environ. Stud.* **2005**, *14*, 335–339.
70. Kumar, M.; Sharma, M.P. Selection of potential oils for biodiesel production. *Renew. Sustain. Energy Rev.* **2016**, *56*, 1129–1138. [CrossRef]
71. Gorji, A. Animal renewable waste resource as catalyst in biodiesel production. *J. Biodivers. Environ. Sci.* **2015**, *7*, 36–49.
72. Sharma, Y.C.; Singh, B.; Korstad, J. Latest developments on application of heterogenous basic catalysts for an efficient and eco friendly synthesis of biodiesel: A review. *Fuel* **2011**, *90*, 1309–1324. [CrossRef]
73. Shan, R.; Zhao, C.; Lv, P.; Yuan, H.; Yao, J. Catalytic applications of calcium rich waste materials for biodiesel: Current state and perspectives. *Energy Convers. Manag.* **2016**, *127*, 273–283. [CrossRef]

74. Nasreen, S.; Nafees, M.; Qureshi, L.A.; Asad, M.S.; Sadiq, A.; Ali, S.D. Review of catalytic transesterifcation methods for biodiesel production. *Intechopen* **2015**, *2*, 64.
75. Guldhe, A.; Singh, B.; Mutanda, T.; Permaul, K.; Bux, F. Advances in synthesis of biodiesel via enzyme catalysis: Novel and sustainable approaches. *Renew. Sustain. Energy Rev.* **2015**, *41*, 1447–1464. [CrossRef]
76. Kouzu, M.; Hidaka, J.S. Transesterification of vegetable oil into biodiesel catalyzed by CaO: A review. *Fuel* **2012**, *93*, 1–12. [CrossRef]
77. Fjerbaek, L.; Christensen, K.V.; Norddahl, B. A review of the current state of biodiesel production using enzymatic transesterification. *Biotechnol. Bioeng.* **2009**, *102*, 1298–1315. [CrossRef] [PubMed]
78. Mata, T.M.; Mendes, A.M.; Caetano, N.S.; Martins, A.A. Properties and sustainability of biodiesel from animal fats and fish oil. *Chem. Eng. Trans.* **2014**, *38*, 175–180.
79. Alptekin, E.; Canakci, M.; Sanli, H. Biodiesel production from vegetable oil and waste animal fats in a pilot plant. *Waste Manag.* **2014**, *34*, 2146–2154. [CrossRef]
80. Araújo, B.Q.; Nunes, R.C.D.R.; De Moura, C.V.R.; De Moura, E.M.; Citó, A.M.D.G.L.; Dos Santos Júnior, J.R. Synthesis and characterization of beef tallow biodiesel. *Energy Fuels* **2010**, *24*, 4476–4480. [CrossRef]
81. Mata, T.M.; Cardoso, N.; Ornelas, M.; Neves, S.; Caetano, N.S. Evaluation of two purification methods of biodiesel from pork lard, beef tallow, and chicken fat. *Energy Fuels* **2011**, *25*, 4756–4762. [CrossRef]
82. De Almeida, V.F.; García-Moreno, P.J.; Guadix, A.; Guadix, E.M. Biodiesel production from mixtures of waste fish oil, palm oil and waste frying oil: Optimization of fuel properties. *Fuel Process. Technol.* **2015**, *133*, 152–160. [CrossRef]
83. Janchiv, A.; Oh, Y.; Choi, S. High quality biodiesel production from pork lard by high solvent additive. *ScienceAsia* **2012**, *38*, 95–101. [CrossRef]
84. Fadhil, A.B.; Al-tikrity, E.T.B.; Albadree, M.A. Transesterification of a novel feedstock, *Cyprinus carpio* fish oil: Influence of co-solvent and characterization of biodiesel. *Fuel* **2015**, *162*, 215–223. [CrossRef]
85. Ejikeme, P.M.; Anyaogu, I.D.; Egbuonu, C.A.C.; Eze, V.C. Pig-fat (Lard) derivatives as alternative diesel fuel in compression ignition engines. *J. Pet. Technol. Altern. Fuels* **2013**, *4*, 7–11.
86. Jeong, G.T.; Yang, H.S.; Park, D.H. Optimization of transesterification of animal fat ester using response surface methodology. *Bioresour. Technol.* **2009**, *100*, 25–30. [CrossRef] [PubMed]
87. Fadhil, A.B.; Ali, L.H. Alkaline-catalyzed transesterification of *Silurus triostegus* Heckel fish oil: Optimization of transesterification parameters. *Renew. Energy* **2013**, *60*, 481–488. [CrossRef]
88. Alptekin, E.; Canakci, M.; Sanli, H. Methyl ester production from chicken fat with high FFA. *World Renew. Energy Congr.* **2011**, *1*, 319–326.
89. Fayyazi, E.; Ghobadian, B.; Najafi, G.; Hosseinzadeh, B.; Mamat, R.; Hosseinzadeh, J. An ultrasound-assisted system for the optimization of biodiesel production from chicken fat oil using a genetic algorithm and response surface methodology. *Ultrason. Sonochem.* **2015**, *26*, 312–320. [CrossRef]
90. Liu, S.; Wang, Y.; Oh, J.H.; Herring, J.L. Fast biodiesel production from beef tallow with radio frequency heating. *Renew. Energy* **2011**, *36*, 1003–1007. [CrossRef]
91. Mata, T.M.; Cardoso, N.; Ornelas, M.; Neves, S.; Caetano, N.S. Sustainable production of biodiesel from tallow, lard and poultry fat and its quality evaluation. *Chem. Eng. Trans.* **2010**, *19*, 13–18.
92. Panneerselvam, S.I.; Miranda, L.R. Biodiesel production from mutton. In Proceedings of the 2011 IEEE Conference on Clean Energy and Technology (CET), Kuala Lumpur, Malaysia, 27–29 June 2011; pp. 83–86.
93. Cunha, A.; Feddern, V.; De Prá, M.C.; Higarashi, M.M.; De Abreu, P.G.; Coldebella, A. Synthesis and characterization of ethylic biodiesel from animal fat wastes. *Fuel* **2013**, *105*, 228–234. [CrossRef]
94. Stojković, I.J.; Miladinović, M.R.; Stamenković, O.S.; Banković-Ilić, I.B.; Povrenović, D.S.; Veljković, V.B. Biodiesel production by methanolysis of waste lard from piglet roasting over quicklime. *Fuel* **2016**, *182*, 454–466. [CrossRef]
95. Vafakish, B.; Barari, M. Biodiesel production by transesterification of tallow fat using heterogeneous catalysis. *Kem. Ind.* **2017**, *66*, 47–52. [CrossRef]
96. Soldi, R.A.; Oliveira, A.R.S.; Ramos, L.P.; César-Oliveira, M.A.F. Soybean oil and beef tallow alcoholysis by acid heterogeneous catalysis. *Appl. Catal. A Gen.* **2009**, *361*, 42–48. [CrossRef]
97. Hu, S.; Guan, Y.; Wang, Y.; Han, H. Nano-magnetic catalyst KF/CaO-Fe3O4 for biodiesel production. *Appl. Energy* **2011**, *88*, 2685–2690. [CrossRef]

98. Xu, G.; Cui, X.; Fan, S.; Zhang, B.; Song, H.; Yan, X.; Ma, X.; Kong, D. Optimization of transesterification of beef tallow for biodiesel production. In Proceedings of the 2011 Asia-Pacific Power and Energy Engineering Conference, Wuhan, China, 25–28 March 2011.
99. Atadashi, I.M.; Aroua, M.K.; Aziz, A.A. Biodiesel separation and purification: A review. *Renew. Energy* **2011**, *36*, 437–443. [CrossRef]
100. Alves, M.J.; Nascimento, S.M.; Pereira, I.G.; Martins, M.I.; Cardoso, V.L.; Reis, M. Biodiesel purification using microand ultrafiltration membranes. *Renew. Energy* **2013**, *58*, 15–20. [CrossRef]
101. Berrios, M.; Skelton, R.L. Comparison of purification methods for biodiesel. *Chem. Eng. J.* **2008**, *144*, 459–465. [CrossRef]
102. Atadashi, I.M.; Aroua, M.K.; Aziz, A.R.A.; Sulaiman, N.M.N. Refining technologies for the purification of crude biodiesel. *Appl. Energy* **2011**, *88*, 4239–4251. [CrossRef]
103. Veljković, V.B.; Banković-Ilić, I.B.; Stamenković, O.S. Purification of crude biodiesel obtained by heterogeneously-catalyzed transesterification. *Renew. Sustain. Energy Rev.* **2015**, *49*, 500–516. [CrossRef]
104. Atadashi, I.M. Purification of crude biodiesel using dry washing and membrane technologies. *Alex. Eng. J.* **2015**, *54*, 1265–1272. [CrossRef]
105. Reis, M.H.M.; Cardoso, V.L. Biodiesel production and purification using membrane technology. In *Membrane Technologies for Biorefining*; Elsevier: Amsterdam, The Netherlands, 2016; pp. 289–307.
106. European Committee for Standardization. *European Standard EN 14214: 2012+A1*; European Committee for Standardization: Brussels, Belgium, 2014; pp. 1–21.
107. U.S. Department of Energy. ASTM Biodiesel Specifications. Available online: https://afdc.energy.gov/fuels/biodiesel_specifications.html (accessed on 5 November 2018).
108. Sakthivel, R.; Ramesh, K.; Purnachandran, R.; Mohamed Shameer, P. A review on the properties, performance and emission aspects of the third generation biodiesels. *Renew. Sustain. Energy Rev.* **2018**, *82*, 2970–2992. [CrossRef]

© 2019 by the authors. Licensee MDPI, Basel, Switzerland. This article is an open access article distributed under the terms and conditions of the Creative Commons Attribution (CC BY) license (http://creativecommons.org/licenses/by/4.0/).

Article

Soybean Oil Transesterification for Biodiesel Production with Micro-Structured Calcium Oxide (CaO) from Natural Waste Materials as a Heterogeneous Catalyst

Samuel Santos [1,*], Luís Nobre [2], João Gomes [1,3], Jaime Puna [1,3], Rosa Quinta-Ferreira [4] and João Bordado [1]

1. CERENA–Centro de Recursos Naturais e Ambiente, Instituto Superior Técnico, Universidade de Lisboa, Av. Rovisco Pais, 1, 1049-001 Lisboa, Portugal; jgomes@deq.isel.pt (J.G.); jpuna@deq.isel.pt (J.P.); jcbordado@tecnico.ulisboa.pt (J.B.)
2. CQE–Centro de Química Estrutural, Instituto Superior Técnico, Universidade de Lisboa Av. Rovisco Pais, 1, 1049-001 Lisboa, Portugal; lcnobre@fc.ul.pt
3. Área Departamental de Engenharia Química, Instituto Superior de Engenharia de Lisboa, Instituto Politécnico de Lisboa, R, Conselheiro Emídio Navarro, 1, 1959-007 Lisboa, Portugal
4. Faculdade de Ciências e Tecnologias, Universidade de Coimbra, R. Sílvio Lina s/n, 3030-790 Coimbra, Portugal; rosaqf@eq.uc.pt
* Correspondence: samuelpsantos@tecnico.ulisboa.pt; Tel.: +351-218-417-755

Received: 21 October 2019; Accepted: 6 December 2019; Published: 9 December 2019

Abstract: In this study, micro-structured calcium oxide obtained from the calcination (850 °C for 3 h) of *Gallus gallus domesticus* (chicken) eggshells was used as a catalyst in the transesterification of soybean oil. This catalyst was characterized by Scanning Electron Spectroscopy (SEM) methods. The structure of the obtained CaO showed several agglomerates of white granular solids with a non-regular and unsymmetrical shape. In terms of calcium oxide catalytic activity, three different catalyst loadings (1%wt, 3%wt, and 5%wt) were tested for the same reaction conditions, resulting in transesterification yields of 77.27%wt, 84.53%wt, and 85.83%wt respectively. The results were compared to the current literature, and whilst they were lower, they were promising, allowing us to conclude that the tendency of yield improvement for this reaction, when the size range of catalyst particles is to be reduced to a nano scale, can be verified.

Keywords: biodiesel; calcium oxide; transesterification; eggshell; solid base heterogeneous catalyst; micro- and nano-structured catalysts

---

1. Introduction

Currently, due to the continuous growth of the world's population, there is a high energy demand in both the industrial and domestic sectors, as well as an increase in public awareness about pollution and the overuse of fossil fuels. This has led to a rise in interest regarding research on alternative renewable energy sources [1–3].

Of the most common renewable energy sources for road transportation, i.e., hydrogen, natural gas, syngas, and biofuel, the latter is the most suitable, environmentally-friendly, and the only one which is ready to be used in vehicles equipped with internal combustion engines (ICE). Biodiesel (from the Greek, bio, life + diesel, from Rudolf Diesel) is the world's most famous biofuel. It is a preferred alternative for petrodiesel (diesel from petroleum oil) in ICE, due to its benefits, such as its availability, non-toxicity, and similar cetane-number, as well as the fact that it can be used directly or in blends with conventional diesel without any need for revamping and even improves the diesel fuel lubrification properties [3–5].

In terms of industrial application for biodiesel production, homogeneous catalysts, such as NaOH and KOH, are usually preferred, but their removal is rather complex and sometimes polluting, bringing extra costs to the final product [6–8]. Considering heterogeneous catalysts for the transesterification reaction, calcium oxide (CaO) is a widely-used catalyst due to being cheap, non-corrosive, economically benign, and easy to handle, in addition to having a high basicity compared to homogeneous base catalysts [2]. It can be obtained from natural sources through the calcination of waste egg and oyster shells (~95% $CaCO_3$) at 850 °C for 3 h, exhibiting high activity for the transesterification of soybean oil due to its superior basic strength [9,10].

On the other hand, heterogeneous catalysts are, for the time being, somewhat time consuming, still inefficient, and still present problems related to mass transfer limitations. One solution regarding this problem might be the use of micro- or nano-structured catalysts, as new heterogeneous catalysts [1,11–14]. Using these based CaO catalysts, it would be possible to overcome some of these issues, as they present a higher surface area and catalytic activity, thus allowing a significant improvement in the transesterification efficiency to be achieved, resulting in faster reactions, i.e., shorter reaction times, low reaction temperatures, and lower catalyst loadings.

Nanocatalysts have recently become the focus of recent research. Reddy et al. (2006) [15] showed that a nanocrystalline CaO was an efficient catalyst for producing biodiesel with high yields at room temperature using soybean oil and poultry fat as raw materials. Hu et al. (2011) [16] developed a nanomagnetic solid base catalyst, $KF/CaO-Fe_3O_4$, based on a magnetic Fe_3O_4 core. In a reaction carried out at 65 °C with a methanol/oil molar ratio of 12:1 and a catalyst concentration of 4% weight related to oil, the biodiesel yield exceeded 95% at 3 h of reaction time. Wen et al. (2010) [17] concluded that the solid base catalysts KF/CaO can be used for biodiesel production with a yield of more than 96%. Kaur et al. (2011) [18] prepared a 1.75 Li-CaO (1.75% weight lithium impregnated CaO) catalyst, which, in the optimized conditions for the transesterification of Karanja and Jatropha oils, could achieve over a 99% conversion of oils to fatty acid methyl esters (FAME).

In the present work, the use of CaO from natural sources, in this case, chicken eggshells, which were then grinded, as a catalyst in the transesterification of soybean oil was studied. The obtained particles from the calcination of the calcium carbonate present in the shells were converted into calcium oxide and then analyzed by Scanning Electron Microscopy (SEM) to assess their structure and particle size. This is an intermediate study regarding the use of nano-structured heterogeneous catalysts for the improved obtention of biodiesel. The results for the soybean oil transesterification will be, in the future, used as a benchmark for a comparison with nanocrystalline CaO catalysts that are currently being studied and prepared by this research team using the Supercritical Anti-Solvent (SAS) method [19–23]. This technique consists of solubilizing CaO into conventional liquid solvents. These solvents are then saturated by supercritical CO_2, resulting in the controlled precipitation of nanocrystalline CaO by the anti-solvent effect [24,25]. Therefore, it will be possible to optimize the nanoparticle size by tuning the operational conditions.

2. Materials and Methods

2.1. Materials

Soybean oil was purchased from a local supermarket in Lisbon, Portugal. Methanol was used in the form of laboratory grade (MeOH; >99% pure). *Gallus gallus domesticus* (chicken) eggshells were collected from several households.

2.2. Preparation of the CaO Catalyst

The eggshells were washed several times with boiling water and then left to dry overnight at 100 °C. After that, the shells were grinded and the obtained solids were sieved into a fine powder using a 30 Mesh (<500 µm) strainer.

For benchmark tests, the calcination was performed for 3 h at 850 °C, with a heating rate of 5 °C/min. The calcium carbonate present in these shells was converted into calcium oxide (CaO), as shown in the equation below:

$$CaCO_3(s) \xrightarrow{850\ °C} CaO(s) + CO_2(g). \tag{1}$$

The obtained calcium oxide was then used as a solid base heterogeneous catalyst for the soybean oil transesterification reaction.

2.3. Catalyst Characterization

Scanning Electron Microscopy images were obtained (JEOL 7100F with an Oxford light elements Energy-dispersive spectroscopy (EDS) detector and Electron backscatter diffraction (EBSD) detector) in order to characterize the produced calcium oxide's morphology and particle size.

Dynamic light scattering graphics were also obtained (Microtrac NANO-flex 180° DLS size) to characterize the calcium oxide's particle size distribution.

2.4. Soybean Oil Transesterification

The transesterification of soybean oil was performed using the CaO resulting from the eggshell calcination and further grinding as the catalyst.

The transesterification reaction took place in a 25 mL flask. In total, 5 g of soybean oil was weighted and heated in a water bath to achieve the reaction temperature of methanol reflux (65 °C). Then, 2.24 g of methanol was weighted and placed inside the reaction flask. The amount of methanol used corresponded to a methanol/oil molar ratio of 12:1. These conditions resulted from work previously developed within this research group on the transesterification of triglycerides using calcium-rich heterogeneous catalysts and the optimization studies then performed [26–29].

Three different catalyst loadings were tested: 1%, 3%, and 5% (w/w, oil basis). The catalysts were added to the methanol, and the mixture was stirred at a high velocity rate. When the methanol reflux temperature was reached, the soybean oil was slowly added to the previous mixture.

Tests with different reaction times were also performed, ranging from one to five hours. Then, the mixture was filtered and placed inside a separation funnel to allow separation of the FAME phase from the glycerol phase.

The yield of the transesterification reaction was calculated using the following equation:

$$\text{Biodiesel yield }(\%) = \frac{\text{Measured weight of FAME}}{\text{Theoretical weight of FAME}} \times 100. \tag{2}$$

The theoretical weight of FAME was calculated using the stoichiometry of the transesterification reaction, as shown in Figure 1.

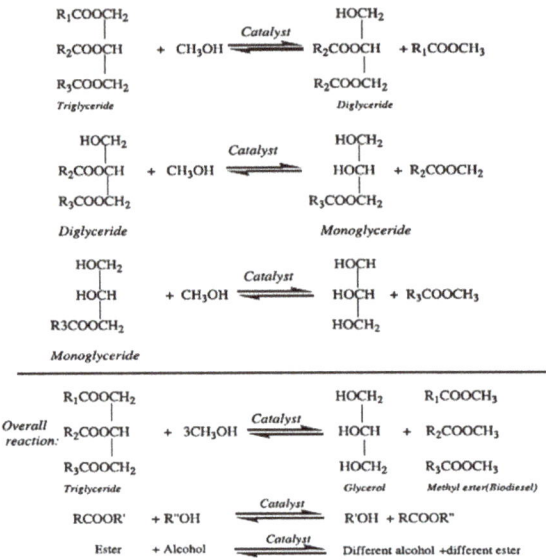

Figure 1. The transesterification of triglycerides and the three steps followed to obtain fatty acid methyl esters (FAME) [4].

3. Results

3.1. Characterization of the Catalyst

3.1.1. SEM Analysis

SEM micrographs of the CaO obtained from the calcination of *Gallus gallus domesticus* (chicken) eggshells were used to identify the morphology of the resulting white powder, as shown in Figure 2.

Figure 2. Scanning Electron Spectroscopy (SEM) microstructures of the CaO obtained from the calcination of eggshells. Eggshells were washed and left to dry overnight at 100 °C, crushed by mortar, and calcined at 850 °C for 3 h. Magnification and bars: (**a**) 500×, 10 μm; (**b**) 5000×, 1 μm; (**c**) 10,000×, 1 μm; (**d**) 30,000×, 100 nm.

3.1.2. Dynamic Light Scattering (DLS) Analysis

The characterization of calcium oxide obtained from the calcination of chicken eggshells was also performed by analyzing the size distribution for the particles. Dynamic light scattering allowed us to assess the size distribution range, which is shown in Figure 3.

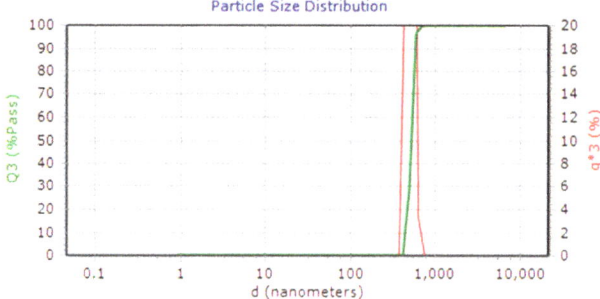

Figure 3. Dynamic light scattering analysis of calcium oxide obtained from eggshells, showing the size range of calcium oxide particles.

3.2. Catalytic Activity

The catalytic activity of the obtained CaO was tested in a bench scale setup at methanol's reflux temperature. The purpose of this experimental work was to assess the optimal reaction conditions for the transesterification of the soybean oil, more specifically, considering the reaction time and catalyst loading. The catalyst loading of 5% and the molar ratio of methanol/oil were previously tested within this research group [26–29]. This was considered a good starting point regarding the evaluation of using calcined eggshells as a heterogeneous catalyst for the transesterification of soybean oil. Additionally, an excess of methanol was necessary, in order to guarantee that the equilibrium shifted towards the products, due to the fact that the transesterification of fatty acids is a reversible reaction, as shown in Figure 3 [5,30].

As shown in Figure 4, for the same molar ratio of methanol/oil (12:1), a catalyst loading of 5% presents the best result in terms of the fatty acid conversion yield, reaching a maximum of 85.83% conversion in only three hours of reaction.

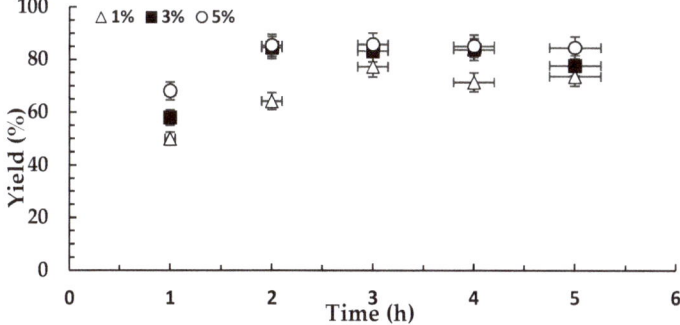

Figure 4. Effect of the reaction time on the FAME yield for three different catalyst loadings.

For the other two catalyst loadings, although the obtained yields are not as high as for the 5% loading, the results for the catalyst loading of 3% exhibit conversion yields in the same range.

A maximum of 84.53% was obtained, for this loading, when the reaction had been running for two hours.

As for the lowest catalyst loading (1%), the highest achievable yield was 77.27%. It is expected that the micronization of this catalyst into nano-structured CaO will result in an increase of its activity, meaning that, possibly, lower amounts will be needed to obtain at least similar FAME yields. In fact, it is expected that FAME yields could be even higher.

4. Discussion

4.1. SEM Analysis

The morphology of the CaO obtained from the calcination of chicken eggshells showed well-developed and defined particles. The structure of the obtained CaO showed several agglomerates of white granular solids, which displayed a non-regular and unsymmetrical shape.

Comparing the obtained morphology with several other studies, it was possible to notice that the observed structure was quite similar. All the micrographs exhibited white non-regular solids where the only noticeable divergence between the various micrographs was the general particle size [31–36].

By analyzing the SEM micrographs for the obtained CaO, it is possible to assume a general particle size smaller than one micron (<1 μm), which, for CaO obtained from calcined eggshells, is the most common size range. Although the particles are smaller than 1 μm, they are not yet in the nano particle range. According to IUPAC (International Union of Pure and Applied Chemistry) the particle's dimensions have to be within the 1–100 nm range in order to be classified as nano [37].

It is expected that, with the implementation of the SAS micronization technique in the future, the CaO particles will become considerably smaller (<100 nm), presenting a regular and symmetrical shape, and thus become nanoparticles.

4.2. DLS Analysis

The size of the obtained CaO particles appears to be distributed within a small region of the nanometric scale.

By analyzing the DLS graphic, it is possible to confirm that that the obtained CaO particle size distribution lies in the narrow gap between 450 and 600 nm, so is indeed smaller than one micron (<1 μm), as mentioned before.

Moreover, it is expected that with the implementation of the SAS technique, CaO nanoparticles with a more homogeneous particle size distribution will be obtained.

4.3. Catalytic Activity

Figure 4 shows the effect of the reaction time on the FAME yields for the three studied catalyst loadings (1%, 3%, and 5%). When comparing this data with that reported from different studies using CaO waste materials as heterogeneous catalysts (Table 1), it is possible to notice that, for different fatty acid feedstocks, in similar reaction conditions, analogous types of catalysts, and catalytic treatments, the obtained results in this study show that, in terms of the transesterification yield, there is still room for improvement using this catalyst. The low yield values could be due to the existence of internal mass transfer limitations, which are related to the hindrance of triglyceride molecules in the CaO micropores and hence, the catalytic activity of the CaO particles is somewhat low. If the size of the CaO particles was reduced to nanoscale material, its surface area would be higher than before, allowing for a lessening of the effect of internal mass transfer limitations, resulting in higher catalytic activity, higher transesterification yields, and shorter reaction times.

Therefore, with the use of nanocatalysts that are currently being prepared, the obtained yields will increase and be in the same size range as that described by Wei et al. (2009).

Table 1. Comparison of yield/conversion to biodiesel using different catalysts prepared from CaO waste materials.

Feedstock	Catalyst	MeOH/Oil Molar Ratio	Catalyst Loading (%wt)	Yield (%) [1]/Conversion (%FAME) [2]	Reference
Soybean oil	Oyster shell (CaO)	6:1	20.0	99.56 [1]	[36]
Soybean oil	Chicken eggshell (CaO)	9:1	3.0	>95.00 [1]	[39]
Used cooked oil	Ostrich/chicken eggshell (CaO)	9:1	1.5	96.00 [2]	[40]
Soybean oil Deodorizer Distillate	Duck eggshell (CaO)	10:1	10.0	94.60 [1]	[41]
Waste cooking oil	Chicken eggshell (CaO) supported char	12:1	10.0	>95.00 [1]	[42]
Sunflower oil	Crab shell (CaO)/Chicken eggshell (CaO)	6:1/9:1	3.0	83.10 [2]/97.75 [2]	[43]
Palm oil	Acid-treated quail eggshell (CaO)	12:1	1.5	>98.00 [2]	[44]
Waste frying oil (WFO)	Commercial CaO	12:1	5.0	87.00 [2] 97.00 [2] 98.00 [2]	[29]
Rapeseed oil					
Soybean oil					
Low FFA (free fatty acid) WFO	Clamshell (*Meretrix meretrix*) (CaO)	12:1	4.0	83.75 [1]/90.13 [2]	[45]

[1] Yield of the transesterification (Equation (1)). [2] Conversion of fatty acid in the oil into fatty acid methyl esters (FAME).

5. Conclusions

CaO waste materials have the potential to be used as micro-structured (and possibly nano-structured) solid base catalysts in the transesterification of triglycerides for the production of biodiesel.

In terms of results for catalyst characterization, SEM micrographs showed agglomerates of granular solids with a non-regular shape, which is common for CaO originating from eggshells.

DLS analysis exhibited particle size distributions in the range of 450 to 600 nm, which means that the obtained solid did not reach a nanoscale. Therefore, there is a need for the application of a micronization technique, such as the Supercritical Anti-Solvent (SAS) method, to achieve particle size distributions below 100 nm (nanoscale).

Regarding the catalyst's activity, it showed somewhat lower transesterification yields when compared with existent literature. These results will be used as a benchmark in the near future, when the transesterification reaction of soybean oil into biodiesel using CaO nanostructured catalysts will be performed and the effect of particle size reduction on the reaction kinetics will be studied. Nevertheless, this study, as an intermediate work, clearly confirms the tendency of yield improvement when the size range of catalyst particles is to be reduced to the nano scale.

Author Contributions: Conceptualization, S.S., L.N., J.G., and J.P.; writing—original draft preparation, S.S. and L.N.; writing—review and editing, J.G. and J.P.; investigation, S.S. and L.N.; supervision, J.B. and R.Q.-F.; project administration, J.B.

Funding: This research was funded by FCT (Fundação para a Ciência e Tecnologia), grant number PD/BD/128450/2017.

Acknowledgments: Support for this work was provided by FCT through PTDC/SEM-ENE/4865/2014.

Conflicts of Interest: The authors declare no conflicts of interest.

References

1. Qiu, F.; Li, Y.; Yang, D.; Li, X.; Sun, P. Heterogeneous Solid Base Nanocatalyst: Preparation, Characterization and Application in Biodiesel Production. *Bioresour. Technol.* **2011**, *102*, 4150–4156. [CrossRef] [PubMed]
2. Safaei-Ghomi, J.; Ghasemzadeh, M.A.; Mehrabi, M. Calcium Oxide Nanoparticles Catalyzed One-Step Multicomponent Synthesis of Highly Substituted Pyridines in Aqueous Ethanol Media. *Sci. Iran.* **2013**, *20*, 549–554. [CrossRef]
3. Meher, L.C.; Vidya Sagar, D.; Naik, S.N. Technical Aspects of Biodiesel Production by Transesterification—A Review. *Renew. Sustain. Energy Rev.* **2006**, *10*, 248–268. [CrossRef]
4. Islam, A.; Taufiq-Yap, Y.H.; Chan, E.S.; Moniruzzaman, M.; Islam, S.; Nabi, M.N. Advances in Solid-Catalytic and Non-Catalytic Technologies for Biodiesel Production. *Energy Convers. Manag.* **2014**, *88*, 1200–1218. [CrossRef]
5. Demirbas, A. Progress and Recent Trends in Biodiesel Fuels. *Energy Convers. Manag.* **2009**, *50*, 14–34. [CrossRef]
6. Singh, A.K.; Fernando, S.D. Transesterification of Soybean Oil Using Heterogeneous Catalysts. *Energy Fuels* **2008**, *22*, 2067–2069. [CrossRef]
7. Liu, C.; Lv, P.; Yuan, Z.; Yan, F.; Luo, W. The Nanometer Magnetic Solid Base Catalyst for Production of Biodiesel. *Renew. Energy* **2010**, *35*, 1531–1536. [CrossRef]
8. Wang, L.; Yang, J. Transesterification of Soybean Oil with Nano-MgO or Not in Supercritical and Subcritical Methanol. *Fuel* **2007**, *86*, 328–333. [CrossRef]
9. Tsai, W.T.; Yang, J.M.; Lai, C.W.; Cheng, Y.H.; Lin, C.C.; Yeh, C.W. Characterization and Adsorption Properties of Eggshells and Eggshell Membrane. *Bioresour. Technol.* **2006**, *97*, 488–493. [CrossRef]
10. Hamester, M.R.R.; Balzer, P.S.; Becker, D. Characterization of Calcium Carbonate Obtained from Oyster and Mussel Shells and Incorporation in Polypropylene. *Mater. Res.* **2012**, *15*, 204–208. [CrossRef]
11. Bet-Moushoul, E.; Farhadi, K.; Mansourpanah, Y.; Nikbakht, A.M.; Molaei, R.; Forough, M. Application of CaO-Based/Au Nanoparticles as Heterogeneous Nanocatalysts in Biodiesel Production. *Fuel* **2016**, *164*, 119–127. [CrossRef]

12. Pandit, P.R.; Fulekar, M.H. Egg Shell Waste as Heterogeneous Nanocatalyst for Biodiesel Production: Optimized by Response Surface Methodology. *J. Environ. Manag.* **2017**, *198*, 319–329. [CrossRef] [PubMed]
13. Baskar, G.; Aiswarya, R. Trends in Catalytic Production of Biodiesel from Various Feedstocks. *Renew. Sustain. Energy Rev.* **2016**, *57*, 496–504. [CrossRef]
14. Chaturvedi, S.; Dave, P.N.; Shah, N.K. Applications of Nano-Catalyst in New Era. *J. Saudi Chem. Soc.* **2012**, *16*, 307–325. [CrossRef]
15. Anr, R.; Saleh, A.A.; Islam, M.S.; Hamdan, S.; Maleque, M.A. Biodiesel Production from Crude Jatropha Oil Using a Highly Active Heterogeneous Nanocatalyst by Optimizing Transesterification Reaction Parameters. *Energy Fuels* **2016**, *30*, 334–343. [CrossRef]
16. Hu, S.; Guan, Y.; Wang, Y.; Han, H. Nano-Magnetic Catalyst KF/CaO–Fe$_3$O$_4$ for Biodiesel Production. *Appl. Energy* **2011**, *88*, 2685–2690. [CrossRef]
17. Wen, L.; Wang, Y.; Lu, D.; Hu, S.; Han, H. Preparation of KF/CaO Nanocatalyst and Its Application in Biodiesel Production from Chinese Tallow Seed Oil. *Fuel* **2010**, *89*, 2267–2271. [CrossRef]
18. Kaur, M.; Ali, A. Lithium Ion Impregnated Calcium Oxide as Nano Catalyst for the Biodiesel Production from Karanja and Jatropha Oils. *Renew. Energy* **2011**, *36*, 2866–2871. [CrossRef]
19. Lee, L.Y.; Wang, C.H.; Smith, K.A. Supercritical Antisolvent Production of Biodegradable Micro- and Nanoparticles for Controlled Delivery of Paclitaxel. *J. Control. Release* **2008**, *125*, 96–106. [CrossRef]
20. Chattopadhyay, P.; Gupta, R.B. Production of Griseofulvin Nanoparticles Using Supercritical CO$_2$ Antisolvent with Enhanced Mass Transfer. *Int. J. Pharm.* **2001**, *228*, 19–31. [CrossRef]
21. Kim, M.-S.; Jin, S.-J.; Kim, J.-S.; Park, H.J.; Song, H.-S.; Neubert, R.H.H.; Hwang, S.-J. Preparation, Characterization and in Vivo Evaluation of Amorphous Atorvastatin Calcium Nanoparticles Using Supercritical Antisolvent (SAS) Process. *Eur. J. Pharm. Biopharm.* **2008**, *69*, 454–465. [CrossRef] [PubMed]
22. Reverchon, E.; Della Porta, G.; Sannino, D.; Ciambelli, P. Supercritical Antisolvent Precipitation of Nanoparticles of a Zinc Oxide Precursor. *Powder Technol.* **1999**, *102*, 127–134. [CrossRef]
23. Tavares Cardoso, M.A.; Antunes, S.; van Keulen, F.; Ferreira, B.S.; Geraldes, A.; Cabral, J.M.S.; Palavra, A.M.F. Supercritical Antisolvent Micronisation of Synthetic All-Trans-β-Carotene with Tetrahydrofuran as Solvent and Carbon Dioxide as Antisolvent. *J. Chem. Technol. Biotechnol.* **2009**, *84*, 215–222. [CrossRef]
24. Reverchon, E.; De Marco, I.; Torino, E. Nanoparticles Production by Supercritical Antisolvent Precipitation: A General Interpretation. *J. Supercrit. Fluids* **2007**, *43*, 126–138. [CrossRef]
25. Fahim, T.K.; Zaidul, I.S.M.; Abu Bakar, M.R.; Salim, U.M.; Awang, M.B.; Sahena, F.; Jalal, K.C.A.; Sharif, K.M.; Sohrab, M.H. Particle Formation and Micronization Using Non-Conventional Techniques—Review. *Chem. Eng. Process. Process Intensif.* **2014**, *86*, 47–52. [CrossRef]
26. Catarino, M.; Ramos, M.; Dias, A.P.S.; Santos, M.T.; Puna, J.F.; Gomes, J.F. Calcium Rich Food Wastes Based Catalysts for Biodiesel Production. *Waste Biomass Valoriz.* **2017**, *8*, 1699–1707. [CrossRef]
27. Soares Dias, A.P.; Puna, J.; Neiva Correia, M.J.; Nogueira, I.; Gomes, J.; Bordado, J. Effect of the Oil Acidity on the Methanolysis Performances of Lime Catalyst Biodiesel from Waste Frying Oils (WFO). *Fuel Process. Technol.* **2013**, *116*, 94–100. [CrossRef]
28. Puna, J.F.; Gomes, J.F.; Bordado, J.C.; Correia, M.J.N.; Dias, A.P.S. Biodiesel Production over Lithium Modified Lime Catalysts: Activity and Deactivation. *Appl. Catal. A Gen.* **2014**, *470*, 451–457. [CrossRef]
29. Puna, J.F.; Correia, M.J.N.; Dias, A.P.S.; Gomes, J.; Bordado, J. Biodiesel Production from Waste Frying Oils over Lime Catalysts. *React. Kinet. Mech. Catal.* **2013**, *109*, 405–415. [CrossRef]
30. Ilgen, O. Dolomite as a Heterogeneous Catalyst for Transesterification of Canola Oil. *Fuel Process. Technol.* **2011**, *92*, 452–455. [CrossRef]
31. Laca, A.; Laca, A.; Díaz, M. Eggshell Waste as Catalyst: A Review. *J. Environ. Manag.* **2017**, *197*, 351–359. [CrossRef] [PubMed]
32. Badrul, H.M.; Rahmat, N.; Steven, S.; Syarifah, F.; Shelly, W.; Agung, P.F. Synthesis and Characterization of Nano Calcium Oxide from Eggshell to Be Catalyst of Biodiesel Waste Oil. *Int. Energy Conf.* **2014**, 340–345.
33. Islam, A.; Taufiq-Yap, Y.H.; Chu, C.M.; Chan, E.S.; Ravindra, P. Studies on Design of Heterogeneous Catalysts for Biodiesel Production. *Process Saf. Environ. Prot.* **2013**, *91*, 131–144. [CrossRef]
34. Gupta, J.; Agarwal, M. Preparation and Characterizaton of CaO Nanoparticle for Biodiesel Production. *AIP Conf. Proc.* **2016**, *1724*. [CrossRef]
35. Viriya-Empikul, N.; Krasae, P.; Nualpaeng, W.; Yoosuk, B.; Faungnawakij, K. Biodiesel Production over Ca-Based Solid Catalysts Derived from Industrial Wastes. *Fuel* **2012**, *92*, 239–244. [CrossRef]

36. Khemthong, P.; Luadthong, C.; Nualpaeng, W.; Changsuwan, P.; Tongprem, P.; Viriya-Empikul, N.; Faungnawakij, K. Industrial Eggshell Wastes as the Heterogeneous Catalysts for Microwave-Assisted Biodiesel Production. *Catal. Today* **2012**, *190*, 112–116. [CrossRef]
37. Lee, B.K.; Yun, Y.H.; Park, K. Smart Nanoparticles for Drug Delivery: Boundaries and Opportunities. *Chem. Eng. Sci.* **2015**, *125*, 158–164. [CrossRef]
38. Nakatani, N.; Takamori, H.; Takeda, K.; Sakugawa, H. Transesterification of Soybean Oil Using Combusted Oyster Shell Waste as a Catalyst. *Bioresour. Technol.* **2009**, *100*, 1510–1513. [CrossRef]
39. Wei, Z.; Xu, C.; Li, B. Application of Waste Eggshell as Low-Cost Solid Catalyst for Biodiesel Production. *Bioresour. Technol.* **2009**, *100*, 2883–2885. [CrossRef]
40. Tan, Y.H.; Abdullah, M.O.; Nolasco-Hipolito, C.; Taufiq-Yap, Y.H. Waste Ostrich- and Chicken-Eggshells as Heterogeneous Base Catalyst for Biodiesel Production from Used Cooking Oil: Catalyst Characterization and Biodiesel Yield Performance. *Appl. Energy* **2015**, *160*, 58–70. [CrossRef]
41. Yin, X.; Duan, X.; You, Q.; Dai, C.; Tan, Z.; Zhu, X. Biodiesel Production from Soybean Oil Deodorizer Distillate Usingcalcined Duck Eggshell as Catalyst. *Energy Convers. Manag.* **2016**, *112*, 199–207. [CrossRef]
42. Gollakota, A.R.K.; Volli, V.; Shu, C.-M. Transesterification of Waste Cooking Oil Using Pyrolysis Residue Supported Eggshell Catalyst. *Sci. Total Environ.* **2019**, *661*, 316–325. [CrossRef] [PubMed]
43. Correia, L.M.; Saboya, R.M.A.; de Sousa Campelo, N.; Cecilia, J.A.; Rodríguez-Castellón, E.; Cavalcante, C.L.; Vieira, R.S. Characterization of Calcium Oxide Catalysts from Natural Sources and Their Application in the Transesterification of Sunflower Oil. *Bioresour. Technol.* **2014**, *151*, 207–213. [CrossRef] [PubMed]
44. Cho, Y.B.; Seo, G. High Activity of Acid-Treated Quail Eggshell Catalysts in the Transesterification of Palm Oil with Methanol. *Bioresour. Technol.* **2010**, *101*, 8515–8519. [CrossRef] [PubMed]
45. Nair, P.; Singh, B.; Upadhyay, S.N.; Sharma, Y.C. Synthesis of Biodiesel from Low FFA Waste Frying Oil Using Calcium Oxide Derived from Mereterix Mereterix as a Heterogeneous Catalyst. *J. Clean. Prod.* **2012**, *29*, 82–90. [CrossRef]

© 2019 by the authors. Licensee MDPI, Basel, Switzerland. This article is an open access article distributed under the terms and conditions of the Creative Commons Attribution (CC BY) license (http://creativecommons.org/licenses/by/4.0/).

MDPI
St. Alban-Anlage 66
4052 Basel
Switzerland
Tel. +41 61 683 77 34
Fax +41 61 302 89 18
www.mdpi.com

Energies Editorial Office
E-mail: energies@mdpi.com
www.mdpi.com/journal/energies

www.ingramcontent.com/pod-product-compliance
Lightning Source LLC
LaVergne TN
LVHW071957080526
838202LV00064B/6772